Data Scheduling and Transmission Strategies in Asymmetric Telecommunication Environments

T0172967

Data Scheduling and Transmission Strategies in Asymmetric Telecommunication Environments

Navrati Saxena
Abhishek Roy

Auerbach Publications
Taylor & Francis Group
Boca Raton   New York

Auerbach Publications is an imprint of the
Taylor & Francis Group, an **informa** business

Auerbach Publications
Taylor & Francis Group
6000 Broken Sound Parkway NW, Suite 300
Boca Raton, FL 33487-2742

© 2008 by Taylor & Francis Group, LLC
Auerbach is an imprint of Taylor & Francis Group, an Informa business

No claim to original U.S. Government works
Printed in the United States of America on acid-free paper
10 9 8 7 6 5 4 3 2 1

International Standard Book Number-13: 978-1-4200-4655-7 (Hardcover)

Library of Congress Cataloging-in-Publication Data

Roy, Abhishek.
    Data scheduling and transmission strategies in asymmetric
telecommunications environments / Abhishek Roy and Navrati Saxena.
        p. cm.
    Includes bibliographical references and index.
    ISBN 978-1-4200-4655-7 (alk. paper)
    1. Telecommunication. I. Saxena, Navrati. II. Title.

TK5101.R69 2007
621.382--dc22   2007017684

Visit the Taylor & Francis Web site at
http://www.taylorandfrancis.com

and the Auerbach Web site at
http://www.auerbach-publications.com

Contents

Preface..ix

Acknowledgments...xi

Authors..xiii

Chapter 1

Introduction..1

1.1   Asymmetric Communication Environments...............................1

1.2   Unicast Versus Broadcast..2

1.3   Push Scheduling Systems..3

1.4   Pull Scheduling Systems...5

1.5   Disadvantages: Push and Pull Systems....................................6

1.6   Hybrid Scheduling Systems...7

1.7   Clients' Impatience...9

1.8   Service Classification and Differentiated QoS............................9

1.9   Multichannel Scheduling..11

1.10 Contribution and Scope of the Work.....................................12

1.11 Organization of the Book...14

Chapter 2

Related Work in Push-Pull Scheduling...15

2.1   Push-Based Systems..15

    2.1.1   Broadcast Disks for Asymmetric Communication................16

    2.1.2   Paging in Broadcast Disks..18

    2.1.3   Polynomial Approximation Scheme for Data Broadcast..........18

    2.1.4   Packet Fair Scheduling..20

    2.1.5   Broadcasting Multiple Data Items................................22

    2.1.6   Broadcasting Data Items with Dependencies....................22

    2.1.7   Broadcast Schedule with Polynomial Cost Functions............23

    2.1.8   Jitter Approximation Strategies in Periodic Scheduling..........23

    2.1.9   Dynamic Levelling for Adaptive Data Broadcasting..............24

2.2   Pull-Based Systems...24

    2.2.1   On-Demand Data Dissemination..................................25

    2.2.2   RxW Scheduling...26

    2.2.3   Data Staging for On-Demand Broadcast..........................27

    2.2.4   Pull Scheduling with Timing Constraints.........................27

    2.2.5   Scheduling with Largest Delay Cost First.........................28

2.3   Both Push and Pull..28

    2.3.1   Lazy Data Request for On-Demand Broadcasting.................29

2.4   Hybrid Scheduling...29

    2.4.1   Balancing Push and Pull...30

    2.4.2   On-Demand Broadcast for Efficient Data Dissemination.........30

    2.4.3   Channel Allocation for Data Dissemination.......................31

	2.4.4	Wireless Hierarchical Data Dissemination System	31
	2.4.5	Adaptive Hybrid Data Delivery	32
	2.4.6	Adaptive Realtime Bandwidth Allocation	32
	2.4.7	Adaptive Dissemination in Time-Critical Environments	33
	2.4.8	Adaptive Scheduling with Loan-Based Feedback Control	33
	2.4.9	Framework for Scalable Dissemination-Based Systems	34
	2.4.10	Guaranteed Consistency and Currency in Read-Only Data	34
	2.4.11	Broadcast in Wireless Networks with User Retrials	35
2.5	Summary		35

Chapter 3
Hybrid Push-Pull Scheduling 37

3.1	Hybrid Scheduling for Unit-Length Items	37
	3.1.1 Assumptions and Motivations	37
	3.1.2 The Basic Hybrid Push-Pull Algorithm	41
3.2	Simulation Experiments	43
3.3	Dynamic Hybrid Scheduling with Heterogeneous Items	45
	3.3.1 Heterogeneous Hybrid Scheduling Algorithm	48
	3.3.2 Modeling the System	50
	3.3.2.1 Minimal Expected Waiting Time	51
	3.3.2.2 Estimation of the Cutoff Value	56
3.4	Experimental Results	58
3.5	Summary	63

Chapter 4
Adaptive Push-Pull Algorithm with Performance Guarantee 65

4.1	Adaptive Dynamic Hybrid Scheduling Algorithm	65
	4.1.1 Analytical Underpinnings	66
	4.1.2 Simulation Experiments	69
4.2	Performance Guarantee in Hybrid Scheduling	71
4.3	Summary	74

Chapter 5
Hybrid Scheduling with Client's Impatience 75

5.1	Hybrid Scheduling Algorithm	75
	5.1.1 Hybrid Scheduling with Clients' Departure	77
	5.1.2 Hybrid Scheduling with Anomalies	77
5.2	Performance Modeling and Analysis	80
	5.2.1 Assumptions	80
	5.2.2 Client's Departure from the System	81
	5.2.3 Anomalies from Spurious Requests	85
5.3	Simulation Experiments	87
	5.3.1 Hybrid Scheduling with Client's Departure	88
	5.3.2 Hybrid Scheduling with Anomalies	91
5.4	Summary	95

Chapter 6
Dynamic Hybrid Scheduling with Request Repetition 97
6.1   Repeat Attempt Hybrid Scheduling Scheme 97
6.2   Performance Analysis of the Hybrid Repeat Attempt System 99
6.3   Simulation Experiments 103
6.4   Summary 106

Chapter 7
Service Classification in Hybrid Scheduling for Differentiated QoS 107
7.1   Hybrid Scheduling with Service Classification 107
7.2   Delay and Blocking in Differentiated QoS 110
       7.2.1   Average Number of Elements in the System 110
       7.2.2   Priority-Based Service Classification 111
                7.2.2.1   Delay Estimation for Two Different Service Classes 112
                7.2.2.2   Effect of Multiple Service Classes 114
7.3   Simulation Experiments 116
       7.3.1   Assumptions 116
       7.3.2   Results 117
7.4   Summary 122

Chapter 8
Online Hybrid Scheduling over Multiple Channels 123
8.1   Preliminaries: Definitions and Metrics 123
8.2   A New Multichannel Hybrid Scheduling 126
       8.2.1   Balanced K Channel Allocation with Flat Broadcast
                Per Channel 127
       8.2.2   Online Balanced K Channel Allocation with Hybrid
                Broadcast Per Channel 128
8.3   Simulation Results 131
       8.3.1   Results 131
8.4   Summary 135

Chapter 9
Conclusions and Future Works 137

References 139

Index 143

Preface

The book is focused exclusively on the various data scheduling and transmission strategies (protocols and algorithms) on telecommunication environments. Comparative advantages and disadvantages of all strategies will be discussed, together with practical consideration and mathematical reasoning.

Background

Although more than a century has passed since the first broadcast of radio signals, the wide popularity of data transmission is relatively recent. The proliferation of Internet and wireless communications from the previous decade has significantly contributed to this upsurge. It begins with basic Web-browsing and e-mail, and expands to include e-commerce, online gaming, and streaming video applications.

Significance

Data broadcasting and scheduling is crucial to all present and future communication systems. With the advancement of communication technologies, data scheduling and broadcasting must be more efficient to deliver information to the target clients in a precise amount of time. For systems (like wireless), which possess asymmetry and uncertainty in communications, data scheduling and broadcasting pose significant challenges. This book presents an in-depth discussion of efficient data scheduling and transmission technologies developed to meet and overcome those challenges.

Audience

1. Universities performing active research in telecommunications. We hope Ph.D. students and professors in particular will benefit.
2. Telecommunication industries (primarily wireless) and research labs performing R&D activities in the area of data scheduling and transmission.

Acknowledgments

Many different people provided help, support, and input to make this book a reality. Professor Maria Cristina Pinotti has always been an enthusiastic advisor, providing encouragement, insight, and a valuable big-picture perspective. She has been a shining catalyst for this work. Additionally, she provided great moral support. We have been fortunate and privileged to work with her.

Thanks to Richard O'Hanley, editor, Auerbach Publications, Taylor & Francis Group, and our editors for their patience and support. We thank the countless people who contributed to this book with informal reviews, suggestions, and improvements. While this is undoubtedly not a complete list, we would like to mention the help of Professor Sajal K. Das and Professor Kalyan Basu.

Finally, we would like to thank our parents, Sweta and Asoke Kumar Roy, and Dr. Vijay Laxmi and Dr. Ashok Kumar Saxena for their love and sincere encouragement. This work would not have been possible without the constant emotional and moral support of our sister, Manu Srivastava and brother-in-law, Ashish Srivastava. They are our last teachers and inspired us with a love of learning and a natural curiosity.

Authors

Navrati Saxena is currently working as a research professor in the electronics and computer engineering department of Sungkyunkwan University, South Korea. During 2005–2006 she worked as an assistant professor and director of research in next generation networks at Amity University, India. Prior to that, in 2003–2005 Professor Saxena worked as a visiting researcher in the computer science and engineering department, University of Texas at Arlington. She completed her Ph.D. from the department of information and telecommunication, University of Trento, Italy. Professor Saxena was also the recipient of Best M.S. and Best B.S. awards from Agra University and Kanpur University, India, respectively. Her prime research interests involve wireless sensor networks, $4G$ wireless, and ubiquitous/smart environments. She has published more than 20 international conference and several international journal articles.
Email:navrati@skku.edu

Abhishek Roy is currently working as an assistant manager in the WiBro (WiMAX) System Laboratory, Telecommunication Network R&D, Samsung Electronics, South Korea. He has five years experience in different aspects of research and development in wireless telecommunication. During 2005 and 2006 he worked as a senior engineer in the Wireless LAN division of Conexant Systems, India. In 2004 and 2005 he was in the systems engineering division of NORTEL Networks, USA. Abhishek received his M.S. degree (with a Best M.S. Research award) in 2002 from the University of Texas at Arlington and a B.E. degree in 2000 from Jadavpur University, Calcutta, in computer science and engineering. His research interests include mobility management, voice over IP (VoIP), and performance modeling of mobile and ubiquitous systems, like WLANs, WiMAX (WiBro), and smart environments. He has published over 15 international conference and 7 international journal articles.
Email:abhishek.roy@samsung.com

1 Introduction

With the immense popularity of the Web, the world is witnessing an unprecedented demand for data services. With the ability to interconnect computers through high-speed bandwidth connections, wireless connections are increasing day by day, and proportionately a new class of services based on data dissemination is also increasing. The major concern of these applications is to deliver data to a very large and impatient population of clients with minimum delay. Many known researchers have investigated this problem of scheduling and broadcasting data for decades, but its optimal solution still remains an open issue.

The recent advancements and ever increasing growth in Web technologies have resulted in the need for efficient scheduling and data transmission strategies. The emergence of wireless communication systems have also added a new dimension to this problem by providing constraints over the low bandwidth upstream communication channels. While today's wireless networks offer voice services and Web browsing capabilities, the actual essence of future generation (3G and Beyond 3G) wireless systems lie in efficient data services. Guaranteeing precise quality of service (QoS), such as the expected access time or delay, bandwidth and blocking are perhaps the most salient features of such data services. To extract the best performance and efficiency of a data transmission scheme, one needs a scalable and efficient data transmission technology.

1.1 ASYMMETRIC COMMUNICATION ENVIRONMENTS

First we will study the environment under consideration. As shown in Figure 1.1, in such data transmission systems, there often exists asymmetry; thus leading to an asymmetric communication environment. This asymmetry may arise because of any of the following factors:

1. As shown in Figure 1.1, the downstream communication capacity (bandwidth from server to client) may be much higher than the upstream communication capacity (bandwidth from client to server).
2. The number of clients is significantly larger than the number of servers.
3. In information retrieval applications, the clients make requests to the server through small request messages that result in the transmission of much larger data items. In other words, asymmetry remains in the size and amount of messages in uplink and downlink transmission.

Thus we assume an asymmetric communication environment for the scope of this book.

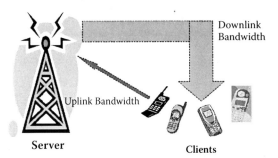

FIGURE 1.1 Bandwidth asymmetry in communication environment.

1.2 UNICAST VERSUS BROADCAST

For such asymmetric environments, in the present day scenario, two types of data dissemination techniques prevail.

One is **unicast**, in which a central service computer replies to each user request in an individual response manner. Unicast servers provide a stream to a single user at a time. This is shown in Figure 1.2.

The other type of data dissemination technique is **one-to-n communication** shown in Figure 1.3. It allows, in one single transmission, multiple clients to receive the data sent by the server. A one-to-n communication could be multicast or broadcast. With multicast, data is sent to a specific subset of clients. Broadcast, in contrast, sends information over a medium on which an unidentified and unbounded set of clients can listen. Multicast utilizes network infrastructure efficiently by requiring the source to send a packet only once, even if it needs to be delivered to a large number of receivers. A wide variety of broadcasting systems prevail in literature, each having different capabilities. A small broadcasting system, for example, is an institutional public address system, which transmits audio messages within an area, e.g., university, shopping mall, or hospital. Another small broadcasting system is low-powered radio or television broadcasting, which transmits audio/video data to a small area. National radio and television broadcasters have nationwide coverage, using satellite systems, cable distribution, and so forth. Satellite radio and television broadcasters can cover even wider areas, such as an entire continent.

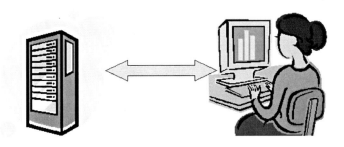

FIGURE 1.2 Unicast communication.

FIGURE 1.3 One-to-n communication.

In one-to-n communication, broadcast is more favorable as it possesses the characteristics of broadcast delivery, heterogeneity, and clairvoyance. Obeying nature's rule that "nothing is perfect," broadcast delivery comes with the limitation of having a bounded best-case performance. This book focuses on various broadcast scheduling algorithms existing in its vast pool of literature. We also propose our new hybrid broadcast scheduling algorithm with many of its variants.

1.3 PUSH SCHEDULING SYSTEMS

Broadly, all broadcast data dissemination applications follow two principles.

In *push-based systems*, clients get required data items by listening to the broadcast channel and capturing the data whenever it goes by. The server broadcasts data items on scheduled time no matter whether the particular item is required or not at that time. The data item to be broadcast next is chosen by a scheduling policy without any clients' intervention.

A push-based system is shown in Figure 1.4. As already mentioned, in a push-based system, the clients continuously monitor a broadcast process from the server and obtain the data items they require, *without making any requests*. Thus, the average waiting time of the clients becomes half of the broadcast cycle. For unit length data items, this result boils down to the half of the total number of items present in the system. The broadcast schedule can be determined online, using a flat round-robin scheme or offline using a packet fair scheduling (PFS) scheme, discussed in the next chapter. Many push systems have been installed. Figure 1.5 shows an example of a real time push system.

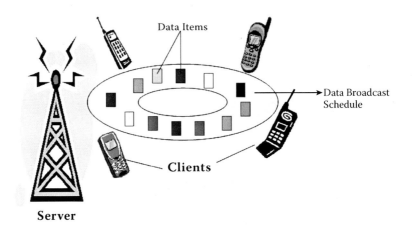

FIGURE 1.4   Push broadcast system.

FIGURE 1.5   An example of a push broadcast system.

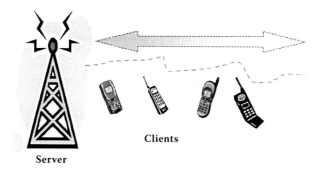

FIGURE 1.6 Pull transmission system.

1.4 PULL SCHEDULING SYSTEMS

In contrast to a push scheduling system, in a *pull-based* system, Figure 1.6, the clients initiate the data transfer by sending requests *on demand*, which the server schedules to satisfy. The server accumulates the client's requests for less-popular items in the pull queue. Subsequently, an item from the pull queue is selected depending on specific selection criteria. This selection criteria depends on the specification and objective of the system. Most request first (MRF), stretch-optimal, priority, or a combination of these techniques is often used. Figure 1.7 shows an example of a pull server.

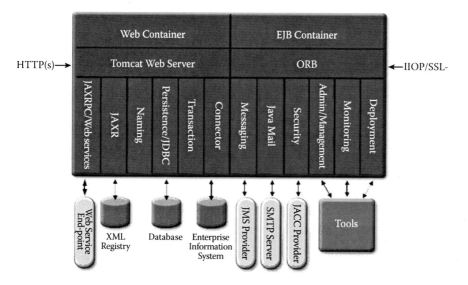

FIGURE 1.7 Example of pull server.

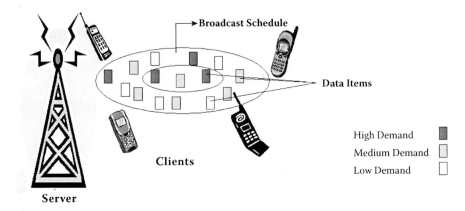

FIGURE 1.8 Disadvantage of push-based system.

1.5 DISADVANTAGES: PUSH AND PULL SYSTEMS

Obeying nature's rule that "nothing is prefect," neither push nor pull alone can obtain the best-case performance. Both push and pull scheduling schemes have their own advantages and disadvantages. Push-based algorithms suffer when the number of data items in the system is large because then the broadcast cycle is long too, and hence the expected access time of the system is also large. For less-demanded items, average access time could be as high as half of the length of the broadcast cycle. A push scheduling is not only affected by the uplink channel constraints, it suffers from wasting resources in downlink wireless channels by repeatedly transmitting the less popular items. Figure 1.8 shows a disadvantage of push-based scheduling where the server broadcasts each item on a scheduled basis, no matter if that item is popular or unpopular among the clients. Thus for a huge set of data items the average length of the push-based broadcast schedule becomes quite high leading to a very high

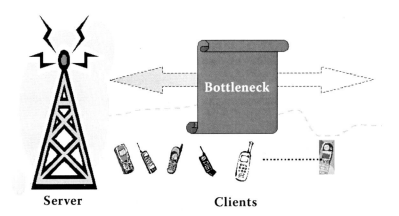

FIGURE 1.9 Disadvantage of a pull-based system.

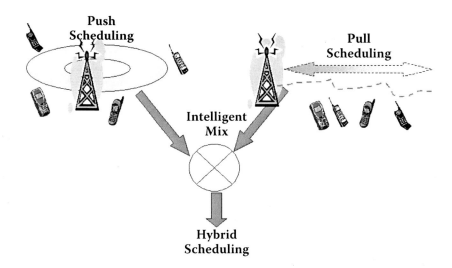

FIGURE 1.10 Intuition for hybrid scheduling.

average waiting time for the clients. Pull-based scheduling is an inefficient scheduling scheme when the system is highly loaded and the data items are very popular. Even though pull-based data dissemination scheme is performed on the basis of explicit clients' requests, such client requests are bounded by the uplink resource (bandwidth) constraints resulting in a bottleneck at the uplink. This is depicted in Figure 1.9.

Hence, neither push nor pull alone can achieve optimal performance [32]. These inefficiencies of push and pull were the main motivation for the hybrid approach of data scheduling. The thought was that if both push and pull scheduling were mixed in such a way that the benefits of both could be availed, then better performance could be gained. So, researchers started with the hybrid approach. This is shown in Figure 1.10.

1.6   HYBRID SCHEDULING SYSTEMS

As shown in Figure 1.11, in a hybrid approach, we divide the set of data items in the server, into two disjoint sets, i.e., access set and request set. The access set contains the *hot data items* that would be sent based on some push scheduling algorithm. The request set contains *cold data items* that would be sent on demand based on some pull scheduling algorithm. Hot data items are the items that are very popular among clients; so it is assumed that they are requested by some clients (among the wide population of the clients) at all times. Cold data items are the data items that are not so popular among the clients. The cutoff point, K as shown in Figure 1.11, separating the items from the push set and the pull set, is chosen in such a way that the overall expected access time is minimized.

A detailed overview of the published research works on wireless data broadcast can be found in [51]. Therefore, the search for efficient hybrid scheduling, which

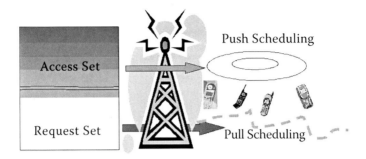

FIGURE 1.11 Essence of hybrid scheduling.

explores the efficiency of both push and pull strategies, continues. Examples of hybrid push-pull systems include the Hughes Network System DirecPC Architecture [23], that uses satellite technology as shown in Figure 1.12 to give a fast, always-on Internet connection; the Cell Broadcast Service (CBS) that enables the delivery of short messages to all the users in a given cell in both GSM and UMTS systems [47]; and the Service Discovery Service in networks of pervasive devices [13].

FIGURE 1.12 Hughes Network's System's DirecPC dish.

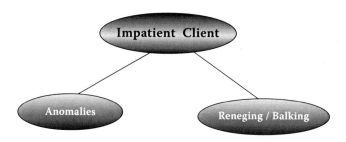

FIGURE 1.13 Effects of impatience.

The general notion of hybrid scheduling lies in dividing the entire set of data items into two parts: *popular items* and *less popular items*. The scheduler pushes the popular data items at regular intervals. It also accumulates the client's requests for less popular items in the pull queue and selects an item depending on the specific selection criteria. A wise selection of the cutoff point, used to segregate the push and pull sets, has the power to reduce the overall expected waiting time of the hybrid system. However, most hybrid scheduling is based on homogeneous (often unit-length) data items. The effect of heterogeneity, with items having different lengths, needs to be considered to get an efficient, hybrid scheduling strategy for asymmetric environments.

1.7 CLIENTS' IMPATIENCE

In practical systems, clients often lose patience while waiting for a particular data item. This results in two-fold effects: (1) As shown in Figure 1.13, the client might get too impatient and leave the system after waiting for a certain time; this is often termed as *balking*. Excessive impatience might result in client's antipathy in joining the system again, which is better known as *reneging*.

The performance of the system is significantly affected by the behavior of the clients. The scheduling and data transmission system needs to consider such impatience resulting in balking and reneging with finite probability. (2) The client may also send multiple requests for the required data item, often known as *anomalies*. Multiple requests by even a single client can increase the access probability of a given item in a dynamic system. In existing scheduling schemes, the server is ignorant of this ambiguous situation and considers the item as more popular, thereby getting a false picture of the system dynamics. Hence, the effects of a client's impatience leading to spurious requests and anomalous system behavior need to be carefully considered and resolved to capture a more accurate, practical behavior of the system.

1.8 SERVICE CLASSIFICATION AND DIFFERENTIATED QoS

Diversification of personal communication systems (PCS) and gradual penetration of wireless Internet have generated the need for differentiated services. The set of clients (customers) in the wireless PCS networks is generally classified into different

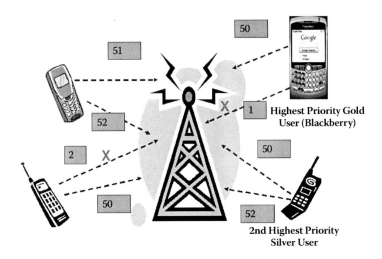

FIGURE 1.14 Different categories of users.

categories based on their importance. Activities of the customers having higher importance have significant impact on the system and the service providers. There could be different categories of users spread randomly among the basic or normal users as shown in Figure 1.14. For example, there could be a higher priority client like a Blackberry user, who pays more to the service provider than the basic user, because he uses the connection for multimedia download and other such uses. We term such a client as a gold user. Similarly, there could be a silver user who uses the connection not only for the voice/data services unlike the basic user, but also for accessing the Internet. Such users would pay more to the service providers than the basic/normal users. The goal of the service providers lies in minimizing the cost associated in the maintenance of the system and reducing the loss incurred from the clients' churn rate. The QoS (delay and blocking) guarantee for different classes of clients should be different, with the clients having maximum importance factor/priority achieving the highest level of QoS guarantee. However, the current cellular systems and their data transmission strategies do not differentiate the QoS among the clients, i.e., the sharing and management of resources do not reflect the importance of the clients. Although most of the service providers support different classes of clients, the QoS support or the service level agreements (SLA) remains the same for all the client classes. Future generation cellular wireless networks will attempt to satisfy the clients with higher importance before the clients having comparatively lower importance. This importance can be determined by the amount of money they have agreed to pay when choosing a particular type of service. Deployment of such differentiated QoS calls for efficient scheduling and data transmission strategies.

However, a close look into the existing hybrid scheduling strategy for wireless systems reveals that most of the scheduling algorithms aim at minimizing the overall average access time of all the clients. We argue that this is not sufficient for future generation cellular wireless systems which will be providing QoS differentiation schemes. The items requested by clients having higher priorities might need to be transmitted in a fast and efficient manner, even if the item has accumulated a smaller number of pending requests. Hence, if a scheduling scheme considers only popularity, the requests of many important (premier) clients may remain unsatisfied, thereby resulting in dissatisfaction of such clients. As the dissatisfaction crosses the tolerance limit, the clients might switch to a different service provider. In the anatomy of today's competitive cellular market this is often termed as *churning*. This churning has adverse impacts on the wireless service providers. The more important the client is, the more adverse is the corresponding effect of churning. Thus, the service providers always want to reduce the rate of churning by satisfying the most important clients first. The data transmission and scheduling strategy for cellular wireless data networks thus must consider not only the probability of data items, but also the priorities of the clients.

1.9   MULTICHANNEL SCHEDULING

To improve the broadcast efficiency in asymmetric communications, one can divide the large bandwidth of the downlink channel into multiple disjoint physical channels. Then, as shown in Figure 1.15, for total push systems, the multiple broadcast problem deals with finding the broadcast schedule on a multichannel environment. The objective is to minimize the *multiple average expected delay* (MAED), that is, the mean of the average expected delay measured over all channels to which a client can afford to listen. To the best of our knowledge, only total push schedules for multiple channels have been proposed so far. Such solutions may either transmit all data items on each channel, or partition the data items into groups and transmit a group per channel. In the former case, MAED can be scaled up to the number of channels that clients can simultaneously listen to by coordinating the solutions for each single channel. In the latter case, clients must afford to listen to all channels, but not necessarily simultaneously. When data items are partitioned among the channels, and the flat schedule is adopted to broadcast the subset of data assigned to each channel, the multiple broadcast problem boils down to the allocation problem introduced in [12, 53]. For such a problem, the solution that minimizes MAED can be found in time polynomial in

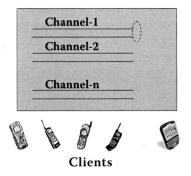

FIGURE 1.15 Downlink divided to multiple channels.

both the size of data and the number of channels [12, 53, 54]. However, the optimal schedule can only be computed offline because it requires inputing the data sorted by decreasing demand probabilities. Moreover, the strategy is not dynamic and the optimal solution has to be recomputed from scratch when the data demand probabilities change. Thus, a need for an efficient online, dynamic, multichannel broadcast scheme arises.

1.10   CONTRIBUTION AND SCOPE OF THE WORK

The prime objective of this book is to develop a framework for new hybrid scheduling strategy for heterogeneous, asymmetric environments. The hybrid scheduling needs to be adaptive and practical enough to be applicable in real-life systems. Subsequently, it should consider the effects of clients' requests as well as their importance to select a particular item for dissemination. More precisely, we can say that the contribution and the scope of the book are as follows:

1. We first propose an ideal hybrid scheduling that effectively combines broadcasting of more popular (i.e., push) data and dissemination upon request for less popular (i.e., pull) data in asymmetric (where asymmetry arises for difference in number of clients and servers) environments. In this approach, the server continuously alternates between one push item and one pull operation. We have assumed an ideal system where the clients send their requests to the server and wait for the necessary data item until they receive it. The data items are initially considered to be of uniform and unit length. At any instant of time, the item to be broadcast is selected by applying a packet fair scheduling (PFS). On the other hand the item to be pulled is the one selected from the pull queue using the most requested first (MRF) scheduling principle.

2. Subsequently, we enhance the proposed hybrid scheduling scheme to incorporate the items having different lengths. Although the push schedule is still based on PFS, the item to be pulled is the one selected from the pull queue using the stretch optimal (i.e, max-request min-service-time first) scheduling principle. We argue that stretch is a more practical and better measure in a heterogeneous system, where items have variable lengths and the difference in item lengths results in the difference in service time of data items. Hence, apart from the client requests accumulated, the system also needs to consider the service time of the items; a larger size item should wait longer than the shorter one. The performance of our hybrid scheduler is analyzed to derive the expected waiting time. The cutoff point between push and pull items is chosen to minimize the overall waiting time of the hybrid system.

3. Subsequently, the hybrid scheduling strategy is further improved so that it does not combine one push and one pull in a static, sequential order. Instead, it combines the push and the pull strategies probabilistically depending on the number of items present and their popularity. In practical systems, the

number of items in a push and a pull set can vary. For a system with more items in the push set (pull set) than the pull set (push set), it is more effective to perform multiple push (pull) operations before one pull (push) operation. We claim that our algorithm is the first work that introduces this concept in a dynamic manner. This has the power to change the push and pull lists in real time and the minimize the overall delay. A strategy for providing a specific performance guarantee, based on the deadline imposed by the clients, is also outlined.

4. In most practical systems, the clients often get impatient while waiting for the designated data item. After a certain tolerance limit, a client may depart from the system, thereby resulting in a drop of access requests. This behavior significantly affects system performance, and needs to be properly addressed. Although an introduction of impatience is investigated in [24], the work considers only pure push scheduling. One major contribution of our work lies in minimizing the overall drop request as well as the expected waiting time.

   There are also ambiguous cases that reflect the false situation of the system. Consider the scenario in which a client gets impatient and sends multiple requests for a single data item to the server. Even if that particular data item is not requested by another client, its access probability becomes higher. In existing systems, the server remains ignorant of this fact and thus considers the item as popular and inserts it into the push or pull set at the expense of some other popular item. In contrast, our work reduces the overall waiting time of the system in the presence of anomalies. More precisely, we develop two different performance models, one to incorporate clients' impatience and the other to address anomaly removal strategy and to analyze the average system behavior (overall expected waiting time) of our new hybrid scheduling mechanism.

5. One major novelty of our work lies in separating the clients into different classes and introducing the concept of a new selection criteria, termed as importance factor, by combining the clients' priority and the *stretch* (i.e, max-request min-service-time) value. The item having the maximum importance factor is selected from the pull queue. We contend that this is a more practical and better measure in the system where different clients have different priorities and the items are of variable lengths. The service providers now provide different service level agreements (SLA), by guaranteeing different levels of resource provisioning to each class of clients. The Quality of Service (QoS) guarantee, in terms of delay and blocking for different classes of clients, now becomes different with clients having the maximum importance factor achieving the highest level QoS guarantee. The performance of our heterogeneous hybrid scheduler is analyzed using suitable priority queues to derive the expected waiting time. The bandwidth of the wireless channels is distributed among the client classes to minimize the request blocking of highest priority clients. The cutoff point used to segregate the push and pull items is efficiently chosen so the overall costs associated in the system are minimized. We contend that the strict

guarantee of differentiated QoS, offered by our system, generates client satisfaction, thereby reducing the churn rate.

6. A new online hybrid solution for the Multiple Broadcast Problem is investigated. The new strategy first partitions the data items among multiple channels in a balanced way. Then, a hybrid push-pull schedule is adopted for each single channel. Clients may request desired data through the uplink and go listen to the channel where the data will be transmitted. In each channel, the push and pull sets are served in an interleaved way: one unit of time is dedicated to an item belonging to the push set and one to an item of the pull set, if there are pending client requests not yet served. The push set is served according to a flat schedule, while the pull set according to the MRF policy. No knowledge is required in advance of the entire data set or of the demand probabilities, and the schedule is designed online.

7. A considerable portion of this book is devoted to the performance analysis of the hybrid scheduling strategies. We have deeply investigated the modeling of scheduling schemes using appropriate tools, like *birth and death process* and *Markov Chain*. The major objective of this performance modeling is to estimate the average behavior of our hybrid scheduling system. Extensive simulation experiments are also performed to corroborate the performance modeling and analysis. We have indicated that a wise selection of a cutoff point to separate push and pull scheduling and consideration of practical aspects (adaptive push-pull, clients' impatience, and service classification) is necessary for a better scheduling strategy with some QoS guarantee.

1.11 ORGANIZATION OF THE BOOK

The book is organized as follows: Chapter 2 introduces the basic push-pull scheduling and offers extensive highlights and comparison of the major existing works in push, pull, and hybrid scheduling. We have introduced our new hybrid scheduling scheme for homogeneous, unit-length items in chapter 3. This chapter also extends the basic hybrid scheduling over heterogeneous (different-length) data items. To make the hybrid scheduling adaptive to the system load, chapter 4 discusses the improvement over this hybrid scheduling and also outlines the basic performance guarantee offered by the hybrid scheduling scheme. The effects of clients' impatience, resulting in their departure from the system and transmission of spurious requests to create an anomalous system behavior and its efficient solution, are discussed in chapter 5. To get a better picture of the clients' retrials and repeated attempts, a different modeling strategy and performance analysis using the multidimensional Markov Chain is developed in chapter 6. The concept of service classification in hybrid scheduling and its effects in providing a differentiated QoS is described in chapter 7. We propose a new hybrid scheduling over multiple channels in chapter 8. Chapter 9 offers conclusions and indications of future research.

2 Related Work in Push-Pull Scheduling

Basically all data transmission mechanisms can be divided into two parts: (1) push-based data broadcasting and (2) pull-based data dissemination. The origin of push-based data broadcasting arises from solving the asymmetry of wireless communication channels. In push-based systems, the server periodically broadcasts a set of data items to all clients, without any client intervention. The clients just listen to the downlink channel to obtain its required data items. The server broadcasts data items according to this schedule. The data item to be broadcast next is chosen again by a scheduling policy without client intervention. Indeed, this saves bandwidth in the resource-constrained uplink wireless channels, but it suffers from wasting resources in downlink wireless channels by repeatedly transmitting the less popular items.

In pull-based systems, a client uses the uplink channel to send an explicit request for a particular data item to the server. The server, in turn, transmits the item to the client. In contrast to the push algorithms, in a pull-based environment, clients make requests through the uplink channel. From time to time, the server considers all the pending requests, makes a schedule, and decides the content of next broadcast. However, pull-based scheduling is inefficient when the system is highly loaded and the data items are very popular.

Many known researchers have long been investigating the problem of data scheduling and broadcasting, but the optimal solution remains an open issue. The main idea is to decrease the expected access time and the expected response time for push- and pull-based systems, respectively. The rest of the chapter offers an extensive review of the major existing works in push, pull, and hybrid scheduling.

2.1 PUSH-BASED SYSTEMS

Push-based broadcast systems explore the downstream communication capacity of wireless channels to periodically broadcast the popular data items. Figure 2.1 demonstrates this basic push-based scheduling principle. The clients present in the system do not need to send an explicit request to the server for any item, thereby saving the scarce upstream channel resources. Instead, the clients simply listen to the server until it receives the desired data item. A wide variety of push-based broadcast scheduling exists in the literature. The vision of efficient push scheduling lies in effectively reducing the overall access time of the data items in an asymmetric communication environment. The concept of broadcast disks, resolving dependencies among different broadcast data items, jitter approximation, and introduction of fair scheduling have contributed to the ultimate realization of this vision. Recent research trends have also addressed the issues related to broadcast of heterogeneous data items and polynomial costs.

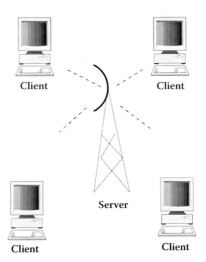

FIGURE 2.1 Push-based communication system.

In this section we take a look into the different major existing push-based broadcast scheduling strategies.

2.1.1 BROADCAST DISKS FOR ASYMMETRIC COMMUNICATION

The concept of the broadcast disk was introduced [1] to explore the downstream channel abundance in asymmetric communication environments. The key idea is that the server broadcasts all the data items to multiple clients. In such a push-based architecture, the broadcast channel essentially becomes a *disk* from which the clients retrieve the data items. The broadcast is created by assigning data items to different disks of varying sizes and speeds. Figure 2.2 represents a schematic diagram of the broadcast disk.

Items stored in faster disks are broadcast more often than items in slower disks. The number of disks, their sizes, and relative speeds can be adjusted to make the broadcast match the data access probabilities. Assuming there is a fixed number of clients with a static access pattern for read-only data, the objective is to construct an efficient broadcast program to satisfy the clients' needs while managing the local data cache of the clients to maximize performance. Intuitively, increasing the broadcast rate of one item decreases the broadcast rate of one or more items. With the increasing skewness of data access probabilities, the flat round-robin broadcast results in degraded performance. Multidisk broadcast programs perform better than skewed broadcasts (subsequent broadcasts of same page clustered together), as illustrated in Figure 2.3. They also aid in prefetching techniques, power savings, and obtaining a suitable periodicity in the broadcast program. The proposed algorithm orders the pages from most popular to least popular. It then partitions the list of the pages into multiple ranges, in which each range contains pages with similar access probabilities. These ranges are termed disks. Now, it chooses the relative frequency of broadcast for

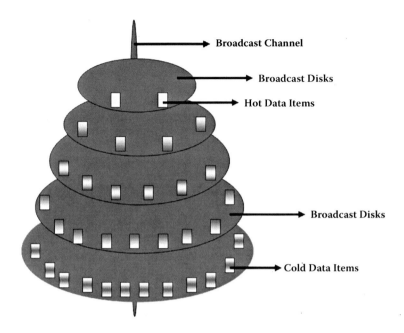

FIGURE 2.2 Broadcast disk.

each disk, which is split into smaller units termed chunks. The broadcast program is created by interleaving the chunks. Thus, the scheme essentially produces a periodic broadcast program with fixed interarrival times per page. Unused broadcast slots are used for transmitting index information, updates, invalidation, or extra broadcast of extremely important pages. Fast disks have more pages than the slower ones.

The major goal in a broadcast disk, for any push-based system, is to minimize the anticipated clients of the system. The expected delay for page requests needs to be estimated for the three different broadcast programs: *flat, skewed,* and *multidisk.* This delay is calculated by multiplying the access probability for each page with the

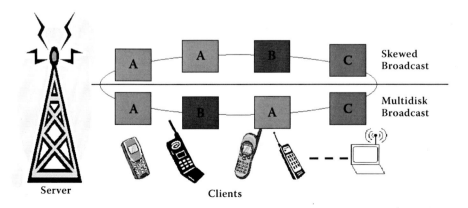

FIGURE 2.3 Broadcast programs.

expected delay for that page and adding the results [1]. Access probability for an item is the measure of the degree of popularity of that item among the clients. Three major points are demonstrated in this aspect:

- For uniform page access probabilities (1/3 each), a flat disk has the best expected performance. This fact demonstrates a fundamental constraint of the broadcast disk paradigm, that due to fixed bandwidth, increasing the broadcast rate of one item must necessarily decrease the broadcast rate of one or more other items.
- As the access probabilities become increasingly skewed, making some items very famous among the clients in comparison to others, the nonflat programs perform increasingly better.
- The multidisk program always performs better than the skewed program.

2.1.2 PAGING IN BROADCAST DISKS

Similar to the concept of virtual memory, paging is also used in broadcast disks to improve the performance. However, a page fault in broadcast disks has variable costs, that depend on the requested page as well as current broadcast state. Also prefetching a page is a natural strategy for performance improvement in broadcast paging. For n data items and a client's cache-size of k, a deterministic algorithm for achieving an $O(n \log k)$ competitiveness in broadcast paging is proposed in [26] Table 2.1. It also points out that in a system without any prefetching, no deterministic algorithm can achieve a competitive ratio better than $\Omega(nk)$. An algorithm is called *lazy* if it moves only when it misses and positions itself on the requested page. Such a lazy algorithm might load a page even if it is not requested as long as no time is spent waiting for that page. A request sequence is *hard* if it faults for every request in the sequence. Comparing the online broadcast paging algorithm $\mathscr{G}$ with *lazy adversaries* reveals that the online algorithm ignores all requests that do not cause any fault and is c-competitive on all such hard sequences. If n and k represent the maximum number of pages in the system and maximum size of client-cache, then for $n = k+1$, there exists a 1-competitive deterministic algorithm for broadcast disk paging. The competitive ratio for a c-competitive deterministic algorithm is $(c-1)n+1$. In fact, without prefetching, no deterministic online algorithm can have a better competitive ratio than $\Omega(nk)$. This result is extended to incorporate randomized algorithms also, with the bound being $\Omega(n \log k)$. A new broadcast paging algorithm, termed as a *Gray algorithm* is proposed. It uses a set of three marks, black, gray, and white, and maintains a mark for each page. For a total number of w requests, the cost on the gray requests in a given segment is at most $O(wn \log k)$. This leads to the result that the amortized cost of an algorithm on white requests is $O(wn \log k)$. Hence, the algorithm is $O(n \log k)$ competitive and can be implemented by keeping track of $O(k)$ gray pages.

2.1.3 POLYNOMIAL APPROXIMATION SCHEME FOR DATA BROADCAST

The first polynomial-time approximation scheme for a data broadcast problem with unit length data items and bounded costs is introduced in [25]. The objective is to minimize the cost of the schedule, where the cost actually consists of expected

TABLE 2.1
Different Scheduling Strategies

No.	Reference	Type	Performance Metric	Adaptability	Special Features
1	[1]	push	response time	no	*LIX, PIX*
					page replacement
2	[3]	pull	stretch value	no	*MAX, AMAX*
					BASE and EDF
					strategy
3	[4, 5]	pull	waiting time	no	scalable, RxW,
					combination of
					MRF and FCFS
4	[6]	pull	waiting time	no	LH-LRU replacement
					opportunistic schedule
5	[30]	hybrid	response time	yes	inaccurate
					data access info.
6	[9]	push	cost (poly. of access time)	no	asymptotic
					lower bound
7	[11]	push	jitter–period tradeoff	no	flexibility
					jitter-period
8	[14]	hybrid	delay	no	real toolkit, scalable,
					LRU cache, information-broker
9	[35]	hybrid	average access time	yes	–
10	[31]	hybrid	access time	no	lazy data
			messaging overhead	no	request
11	[44]	hybrid	throughput	no	data consistency
			abort	no	concurrency
12	[8]	push	access time	no	separating service
					provider entity
13	[33]	push	expected delay	yes	sensitivity with items,
					disks and frequencies
14	[29]	push	access time	no	multiple
			query frequency		data items
15	[21]	push	delay	yes	hierarchical data deliver model
16	[22]	push	message traffic	yes	group info
					loan-based slot-allocation
					feedback control
17	[26]	push	cost	no	$O(n \log k)$ competitive
18	[46]	pull	access time	yes	largest delay
			tuning time		cost first
			failure recovery		(LDFC)
19	[34]	hybrid	average access time	yes	–
20	[7]	push	waiting time	no	file dependency
21	[32]	hybrid	average access time	yes	dynamic popularity
			average completion time		of data items
22	[25]	push	broadcast cost	no	fast, polynomial
				approx. algos.	approach

(*Continued*)

TABLE 2.1 (CONTINUED)
Different Scheduling Strategies

No.	Reference	Type	Performance Metric	Adaptability	Special Features
23	[19, 49]	push	access time	no	packet fair scheduling
24	[46]	push-pull	access time response time	yes	identical push-pull systems
25	[16]	push	requests scheduled missed deadlines	yes	time constraints
26	[27, 28]	push-pull	access time on-demand channels	yes	cost estimation of dynamic scheduling
27	[2]	hybrid	response time	no	scalability issues
28	[15]	pull	deadlines missed	no	aggregated critical requests (ACR)
29	[52]	hybrid	delay	no	EDF, deadline, and batching

response time and broadcast cost of the messages. The basic idea is to form different groups consisting of equivalent messages (i.e., messages having the same cost and probability). Within every group these messages are rearranged in such a manner that they can be scheduled in a cyclic, round-robin fashion. This concept is extended to a generalized case of broadcast scheduling. A randomized algorithm is introduced that rounds the probabilities and costs of messages, and partitions them into three similar groups. Subsequently a greedy technique is also introduced that minimizes the expected cost of the already allocated slots. This greedy schedule is at least as good as the randomized schedule. The period of this greedy technique is bounded in polynomial length.

2.1.4 PACKET FAIR SCHEDULING

The work of Hameed and Vaidya [19, 49] relates the problem of broadcast scheduling with *packet fair scheduling* (PFS) and presents an $O(\log D)$ scheduling algorithm for D number of data items. It introduces the concept of *spacing* between two items as the time taken between two consecutive broadcasts of a particular (same) data item. Figure 2.4 depicts a part of the broadcast cycle in which each item has similar length. On the other hand, Figure 2.5 shows a part of the broadcast cycle in which items are assumed to be of variable length.

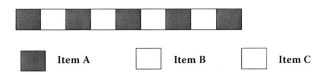

FIGURE 2.4 Packet fair scheduling: items of same length.

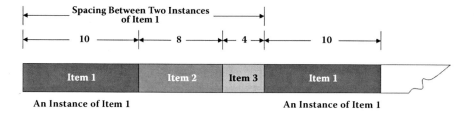

FIGURE 2.5 Packet fair scheduling: items of variable length.

For optimal scheduling, any data item needs to be equally spaced [19, 49]. If l_i and s_i represent the length and spacing of item i, then assuming a Poisson arrival of client requests, the waiting time of any client for that particular item is given by: $t_i = s_i/2$. Now, if p_i represents the probability of item i, then the overall mean access time ($t_{overall}$) is given by $t_{overall} = \sum_{i=1}^{D} p_i t_i = \frac{1}{2} \sum_{i=1}^{D} p_i s_i$. At this point of time one needs a suitable, optimal expression of spacing s_i. If instances of all items are equally spaced, then minimum overall access time is achieved when s_i and $t_{optimal}$ are given by the following equations:

$$s_i = \left[\sum_{j=1}^{D} \sqrt{p_j l_j} \right] \sqrt{\frac{l_i}{p_i}} \qquad (a)$$

$$t_{optimal} = \frac{1}{2} \left[\sum_{i=1}^{D} \sqrt{p_i l_i} \right]^2 \qquad (b) \qquad (2.1)$$

Although equal spacing of data items is not always feasible in practical systems, $t_{optimal}$ provides a lower bound on the overall minimum expected access time. The packet fair scheduling algorithm operates as a switch, connecting many input queues with a single output queue. The objective is to determine which packet will be transmitted from the set of input queues to the output queue. For a specific value ϕ_i, the input queue i should get at least fraction ϕ_i of the output bandwidth. Thus bandwidth is evenly distributed between the input queues. For optimal scheduling the spacing between consecutive instances of same data item i needs to be obtained from Equation 2.1(a). Thus we have,

$$\frac{l_i}{s_i} = \frac{l_i}{\left(\sum_{j=1}^{D} \sqrt{p_j l_j} \right) \sqrt{l_i/p_i}} = \frac{p_i l_i}{\sum_{j=1}^{D} \sqrt{p_j l_j}} \qquad (2.2)$$

The performance of the algorithm can be further improved by using suitable bucketing techniques [49]. However, wireless channels are inherently lossy and error-prone. Thus any practical broadcast scheduling should consider the associated transmission errors. Although error control codes (ECC) help to correct these errors, it is not possible to correct all errors. Any erroneous packet is discarded after reception.

2.1.5 BROADCASTING MULTIPLE DATA ITEMS

Traditional broadcasting schemes, which do not consider the relationship between data items, often increase the average access time to process client requests in this environment. The problem of deciding and scheduling the content of the broadcast channel is found to be NP-hard [29]. Subsequently different greedy heuristics exist for obtaining near optimal solutions.

It is quite clear that deciding the content of the broadcast channels is based on clients' queries. Given a set of queries and a set of equal sized data items, each query accesses a set of data items termed *query data set*. For a given set of data items and queries, the query selection problem is to choose a set of queries that maximizes the total overall frequency of queries, constrained to the number of data items over all queries to be bounded by maximum possible number of data items currently present in the channel. Three different greedy approaches based on (1) highest frequency, (2) highest frequency/size ratio, and (3) highest frequency/size ratio with overlapping are proposed to solve the query selection problem.

The proposed query expansion method sorts the queries according to their corresponding access frequencies and inserts the data items of each query in a greedy manner. Higher frequency queries are given higher preferences for expansion. This basic method is extended to also include the frequencies of data items. To reduce the overall access time, the query set of the overlapping and previously expanded data items are modified by moving the data items to either left-most or right-most positions of the previous schedule. This change makes the data items adjacent to data items of currently expanded query. The moving of queries is performed only if the number of remaining queries benefitting from this operation is larger than the remaining queries suffering from it. On the other hand, the content of the scheduling can be expanded on the basis of data items also. The data items of chosen queries are transformed to a data access graph – a weighted, undirected graph. Each vertex represents a certain data item and each edge represents the two data items belonging to a certain query. The procedure combines two adjacent vertices of the graph into a multivertex. If any vertex has more than one edge to a multivertex, the edges are coalesced into a single edge with the previous edge-weights added to form the total new weight. The procedure is iterated until the graph is left with a single multivertex.

2.1.6 BROADCASTING DATA ITEMS WITH DEPENDENCIES

Research has demonstrated the existence of a simple optimal schedule [7] for two files. Considering all possible combinations of clients from both the classes and accessing the data items of any single or both the classes, the work has shown that for equal length files with no dependencies, any optimal schedule can be partitioned into consistent simple segments, i.e., there exists an optimal simple schedule. The model is also extended to incorporate variable length file sizes. It has been proved that a simple, optimal schedule still exists.

While the objective of a broadcast schedule is to minimize the access cost of a random client, most of the schedules are based on the assumption that access cost is directly proportional to the waiting time. However, in real scenarios the patience of a client is always not necessarily proportional to the waiting time. This makes the broadcast scheduling problem even more challenging by generating polynomial cost functions.

2.1.7 BROADCAST SCHEDULE WITH POLYNOMIAL COST FUNCTIONS

Recent research [10] has shown the formulation of fractional modelling and asymptotically optimal algorithms for designing broadcast schedules having cost functions arbitrary polynomials of a client's waiting time. For any data item $i \in \{1, 2, 3, \ldots, D\}$ with probability p_i a cyclic schedule is a repeated finite segment. For any increasing cost function, the optimal fractional broadcast schedule with minimum expected cost results when successive instances of each data item are equally spaced. For such a model, the optimality is achieved when the frequency of each item satisfies the relation: $frequency_i = \frac{\sqrt{p_i}}{\sum_{j=1}^{D} \sqrt{p_j}}$. The access probabilities of the data items are assumed to obey Zipf distribution, with access skew coefficient θ. Subsequently a random algorithm, a halving algorithm, a fibonacci algorithm, and a greedy algorithm is proposed to obtain the optimal fractional schedule.

1. At each step, the randomized algorithm transmits a page such that expected frequency of each page approaches exact frequency of fractional schedule. For linear (first order) function, a randomized algorithm provides a solution bounded by *twice* the optimal cost. However, the performance of the randomized algorithm deteriorates exponentially for nonlinear function.
2. A halving algorithm attempts to achieve the optimal fractional schedule by rounding off the desired page frequencies to the nearest power of $1/2$. When the desired frequencies are always powers of 2, the strategy achieves the optimal schedule. On the other hand, in the worst case, the actual frequency is always more than $1/2$ the original frequency. For a linear cost model, the halving algorithm results in costs bounded by twice the optimal algorithm.
3. Like a random algorithm, the fibonacci index (golden ratio) algorithm also generates a schedule with the same average frequencies as those of the optimal fractional solution. However, the spacing between two consecutive appearances of same item in the schedule may have three different schedules close to the optimal periods. For a linear cost function, the fibonacci algorithm generates a schedule whose cost is $9/8$ times the cost of optimal fractional model.
4. At every step the greedy algorithm broadcasts the item that will have maximum cost if not broadcasted. A finite schedule is computed and repeated at each iteration of the algorithm. Even with exponential cost function, the greedy approach results in a very near optimal solution.

2.1.8 JITTER APPROXIMATION STRATEGIES IN PERIODIC SCHEDULING

Perfect periodic scheduling broadcasts each item at exact time intervals. This removes the constraint for the client to wait and listen to the server until its desired item is broadcast. Instead, the client now has the flexibility to switch the mobile on exactly when needed, thereby saving the energy of power-constrained hand-held mobile terminals. Jitter is estimated as the difference in spacing between the consecutive occurrences of the same data item. A new algorithm for controlling the jitter in the schedule is proposed. It requires the ratio between any two periods to be a power of 2. The key idea is to evenly spread the schedule over the entire period in a recursive fashion. Idle slots are inserted in the schedule to remove the imperfect balancing. Using a suitable parameter, the algorithm controls the influence of jitter and period approximation. It first constructs a binary tree to create a replica for each job in the instance and associates these replicas with the root of the tree. Each node has exactly two children. In order to ensure low jitter, the strategy uses a total ordering of jobs. This results in reduction of the jitter in data broadcasting.

2.1.9 DYNAMIC LEVELLING FOR ADAPTIVE DATA BROADCASTING

The major problem associated with this research on broadcasting over multiple channels lies in generating hierarchical broadcast programs with a given number of data access frequencies and a number of broadcast disks in a broadcast disk array. The problem of generating hierarchical broadcast programs is first mapped into construction of a channel allocation tree with variant fanout [33]. The depth of the allocation tree corresponds to the number of broadcast disks, and the leaves in the same level actually represent the specific data items. The data items in the fast disks are accessible faster than the data items in slower disks. However, the data access frequencies change over time. The broadcast programs need to dynamically adapt to all such changes.

2.2 PULL-BASED SYSTEMS

While push-based broadcast strategy attempts to reduce the overall expected access time, it suffers from two major disadvantages:

1. The server broadcast does not discriminate between the popular (hot) and nonpopular (cold) items. Thus, the nonpopular (cold) items are also broadcast repeated times in periodic intervals. This results in wastage of valuable bandwidth, because the nonpopular items are required by a handful of clients.
2. On an average, the overall expected access time becomes half of the entire broadcast cycle length. Hence, for a system having very large number of items, some of which are nonpopular, the average waiting time for the popular items also becomes pretty high. In other words, the popular items suffer because of the presence of nonpopular items.

A close look into the scenario reveals that the major reason behind these two problems lies in the absence of clients' explicit role in the scheduling. Indeed, push-based scheduling does not take clients' needs into account. This gives rise to the on-demand pull scheduling.

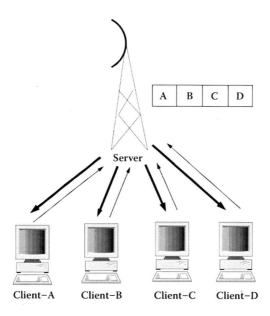

FIGURE 2.6 Pull-based communication system.

Figure 2.6 shows the basic of on-demand pull scheduling. In a pull-based data transmission scheme, the clients explicitly send an uplink request for a particular data item to the server. The server, in turn, processes the requests and transmits the data item over a downlink channel. A wide variety of scheduling principles exist for this pull-based scheduling. Although, MRF provides a low average access time, it suffers from fairness. On the other hand, FCFS, is fair, but suffers from sub-optimality and increased average access time. A combination and modification of these basic scheduling principles give rise to other scheduling strategies, like shortest time first (STF) and lowest waiting time first (LWTF). Subsequently, caching, prefetching, and opportunistic scheduling are also used to improve the performance of on-demand pull-based data transmission. The eventual goal is to satisfy certain timing constraints imposed by real-time communication. In this section we take a look into the major pull-based scheduling strategies.

2.2.1 ON-DEMAND DATA DISSEMINATION

The scheduling problems arising in on-demand broadcast environments for applications with heterogeneous data items are investigated in [3]. A new metric *stretch* is introduced for performance measurement in heterogeneous environments. The primary objective of the proposed algorithm is to optimize the worst case stretch of individual requests. Like other pull-based applications, the clients send explicit requests to the server, and the server transmits the specific item to the client. The transmission unit is a page, a basic fixed-length unit of data transfer between clients and server. The pages are assumed to have self-identifying headers that are delivered in a specific order.

The concept of preemption is used to achieve better scheduling performance. This also aids in implementing the scheduling strategy with less complexity, as most of the nonpreemptive scheduling schemes are NP-hard. The preemption helps avoid the backlog of pending requests when a long job is being serviced. Preemption of an item for a more popular item also has the potential for the improvement of its performance. While response time is the most popular performance measure, it is not a fair measure in heterogeneous systems. Individual requests differ in terms of their service time. The proposed strategy uses the concept of stretch, defined as the ratio of response time of a request to its service time. Stretch explores the intuitive concept that larger jobs should take more service time than smaller jobs. The jobs are classified based on their service times. The average of maximum stretch for each class (AMAX) aids to get the overall picture of the entire system. This classification helps in better under-standing of system performance. This algorithm repeatedly guesses a stretch-value, which immediately yields a *deadline* for each job based on its arrival and service time. Earliest deadline first (EDF) is used to determine if all jobs can be scheduled with a bounded maximum stretch, thereby meeting the respective deadlines. The objective is to use the past access-history to make an intelligent guess of stretch. The current stretch is used to obtain the deadline.

2.2.2 RxW SCHEDULING

The RxW algorithm [4, 5] is proposed to meet these criteria. By making the scheduling decisions based on current request queue state, RxW can adapt to the changes in client population and workload.

The primary idea behind designing the RxW algorithm is to exploit the advantages of both MRF and FCFS. While MRF provides lowest waiting time for popular pages, it suffers from fairness and might lead to starvation of nonpopular requests. On the other hand, FCFS is fair but suffers from higher waiting time. The success of LWF lies in providing more bandwidth to popular pages, while avoiding starvation of nonpopular pages. RxW combines the benefits of both MRF and FCFS in order to provide good performance to both popular and nonpopular items, while ensuring scalability and low overhead. Intuitively, it broadcasts every page having the maximal $R \times W$ values, where R and W are the number of pending requests and time of the *oldest outstanding request* for the particular page. Three different versions of the RxW algorithm are proposed:

1. *Exhaustive RxW*: The exhaustive RxW maintains a structure containing a single entry for each page having outstanding requests. It also maintains R, first arrival time. For any arriving request, a hash look up is performed to get the page. If the request is the first one, then the R is initialized to 1 and first arrival is initialized to current time; otherwise the value of R is incremented. The server selects the page having largest $R \times W$ value.

2. *Pruning Search Space*: This version of the algorithm uses two sorted lists (1) the W-list, ordered by increasing order of first arrival time and (2) the R-list, ordered by decreasing order of R-values. The entries in the W-list are kept fixed until the page is broadcasted. However, the entries in the R-list

are changed during every request arrival. This makes request processing a constant-time operation. The pruning scheme truncates the W-list. The algorithm alternates between two lists, updating the maximum value of $R \times W$ at every iteration.

3. *Approximating RxW*: The approximated, parameterized version of RxW reduces the search space even further at the cost of suboptimal broadcast decision. For highly skewness data items, the maximal $R \times W$ value is obtained at the beginning of the least one of the two lists. Also, the static workload, the average $R \times W$ value of the page chosen to be broadcast converges to a constant.

2.2.3 DATA STAGING FOR ON-DEMAND BROADCAST

The RxW algorithm [4, 5] is extended to incorporate these data staging strategies [6] for improved performance of on-demand data broadcasting. The server maintains a service queue for keeping the outstanding clients' requests. Upon receiving a request, the queue is checked. If the item is already present, then the entry is updated, otherwise, a new entry for that item is created. An ending is also kept to track the items for which an asynchronous fetch request is pending. The limit of this request is constrained by the size of the pending list. The server first checks for completion of any asynchronous fetches in the pending list. The items arriving by fetch operation are broadcasted in the order they were received and the corresponding entries are removed from the pending list. If the pending list is not full, the server operates in normal mode, otherwise, it operates in opportunistic mode. In the normal mode, the server selects an item using RxW algorithm. If the selected item is found in cache, it is broadcast, otherwise an entry to the item is created in the pending list and request for the item is sent to the remote site or secondary/tertiary item. When the server has reached the limit of outstanding requests, the system switches to opportunistic scheduling mode. The algorithm is now restricted to cache-resident pages, having at least one pending request. A single bit in the service queue is kept to check whether the page is cache-resident or not. The algorithm now attempts to find the best available (cache resident) page according to RxW strategy. A new modified LRU replacement scheme, termed as LRU with love/hate hint (LRU-LH) is used to improve the cache replacement strategy. The popular and nonpopular pages are marked as 'love' and 'hate' to put them in the top and bottom of the LRU chain. A page is selected for broadcast if it is encountered on the R-list before the W-list.

2.2.4 PULL SCHEDULING WITH TIMING CONSTRAINTS

An investigation into traditional realtime nonmobile and nonrealtime mobile data transmission strategies is performed in [15]. Subsequently an efficient pull-based scheduling scheme based on aggregated critical requests (ACR) is designed to meet the specific deadline of clients' requests.

In realtime nonmobile environment, the server assigns priorities to transactions based on several strategies, such as, earliest deadline first (EDF) or least slack (LS) first. As the name suggests, in EDF the item with earliest deadline is given the highest priority. On the other hand, in LS the slack time at any instant t is estimated using the expression: $deadline - (t + executionTime - processorTime)$. The transaction is capable of meeting the deadline if the slack time is zero. Although EDF is the best overall strategy, it performs in a very poor manner when the system load is high. In pull-based, mobile nonrealtime strategies, the longest wait first (LWF) often outperforms all other schemes to achieve the minimum waiting time. LWF computes the sum of total time that all pending requests have been waiting for a data item. The database is assumed to consist of a fixed number of uniform pages in which each page fits into one slot. Broadcast time of each page is equal to one slot. Assuming a Poisson arrival rate, the system assigns the slots to particular data items such that the *long term deadline miss ratio* is minimized. At any time slot t this ACR strategy attempts to minimize the deadlines missed during time slot $t + 1$ by transmitting the page with the most deadlines to meet before slot $t + 1$. The waiting requests are kept in the pull queue. The server maintains the number of deadlines to be missed if a page is not transmitted in the next time slot. The requests corresponding to the deadlines are termed as *critical requests* and the server updates the number of critical requests for the data item at every time slot. It chooses the page having the largest number of critical requests to transmit, deletes the requests with missed deadlines, and resets the number of critical requests.

2.2.5   SCHEDULING WITH LARGEST DELAY COST FIRST

While most broadcast scheduling strategies (both adaptive and nonadaptive) attempt to minimize the overall access time, recent researches have been focussed to reduce the overall delay cost [46] in on-demand pull-based data dissemination schemes. The delay cost consists of three different components. Apart from the existing overall access time cost, it also takes the tuning time costs and failure recovery costs into account. Like conventional pull scheduling schemes, in the proposed *largest delay cost first* (LDCF), the clients explicitly send the request for specific data items to the server. However, it does not wait indefinitely for server's response. Instead, the clients use a response time limit (RTL) to indicate the maximum possible time it can wait for server's response. The strategy also considers tuning time costs that correspond to the search for the location of a particular data item in the index. During the entire broadcast period, the strategy receives the new requests and adds them into the request sequence. The data item with largest priority is selected and added to the broadcast period, sorted by the descending order of popularity factor. The index is obtained and the data item is broadcasted. The failure requests are now cleared.

2.3   BOTH PUSH AND PULL

Now, it is clear that both push-based and pull-based scheduling have some specific advantages. Naturally, a wise approach is to explore the advantages of both of these basic data transmission mechanisms. This gives rise to some interesting data transmission schemes, such as lazy data request.

2.3.1 Lazy Data Request for On-Demand Broadcasting

While the prime goal of data broadcasting is reducing the overall access time, most practical systems also need to consider the messaging overhead. On the other hand, the load of the real-time systems often change in a dynamic fashion. Hence, the broadcast system needs to be robust enough to adapt itself online with the system dynamics. The basic motivation behind the lazy data request strategy [31] is to not send the request for the data item but wait. The particular data item might already be broadcasted due to an explicit request by other clients. This will result in saving of message passing in the uplink channel and battery power of the mobile terminal. It proposes a new dynamic bounded waiting strategy that contains two parts in the schedule: index section and data section. The client can use the index section to get a predicted estimate of the item to be broadcasted in the near future. The server chooses a set of data items and the items are batched together for broadcast. The corresponding index section is broadcasted before the transmission of the batch set. The number of items to broadcast in a batch set is determined by a control variable termed selection factor. In the worst case the client tunes at the beginning of a data section and waits until the end of next index section of the next data set.

2.4 HYBRID SCHEDULING

Hybrid approaches that use the flavors of both push-based and pull-based scheduling algorithms in one system appear to be more attractive. The idea, as shown in Figure 2.7, is to exploit an intelligent of both push and pull scheduling, in such a way that the benefits of both could be availed to achieve performance of the system. The key idea of a hybrid approach is to separate the data items into two sets: (1) popular and (2) nonpopular. While the popular data items are broadcast using

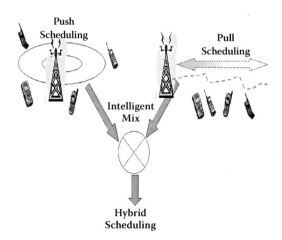

FIGURE 2.7 Motivation of hybrid scheduling.

push-based transmission strategy, the relatively nonpopular items are transmitted using on-demand pull-based scheduling strategy. A suitable balance between push and pull schemes is of utmost importantance at this aspect. A major characteristic of an efficient hybrid scheduling strategy is its adaptiveness. The strategy should be able to change the scheduling decisions online. In this section we will look into the major hybrid scheduling techniques.

2.4.1 BALANCING PUSH AND PULL

The work of Acharya, Franklin, and Zdonik [2] is perhaps the first significant work that effectively explores the advantages of both push- and pull-based data transmission strategies. The work introduces the asymmetry in different factors, like, (1) uplink and downlink channels, (2) clients and server ratio, and (3) amount of data downloaded and uploaded. The proposed strategy considers a capacity-constrained server and multiple clients with uplink channels. It then extends the static, push-based data broadcasting to incorporate pull-based on-demand data transmission schemes for read-only data items. For push-based broadcast scheduling, the proposed algorithm selects the cached page which contains lowest p/x ratio. The pull-based on-demand scheduling is modelled as a point-to-point connection with the server. While the rate of client-requests increases as the number of clients increases, the server only has a maximum number of allowable requests it can handle. The server is capable of interleaving push- and pull-based data items, and options are kept to vary the percentage of slots dedicated for on-demand pull scheduling. The requests are accumulated and kept in the pull queue. The server selects the item in a first-come-first-serve (FIFO) fashion. A threshold parameter is kept to maintain the use of the back channel under certain limits. While *measured client* models a single client, the *virtual client* models the combined effect of all other clients in the system. It maintains a cache holding different pages and waits for certain time units between two consecutive requests. If possible, the requests are serviced from the cache; otherwise they are broadcast or pulled.

2.4.2 ON-DEMAND BROADCAST FOR EFFICIENT DATA DISSEMINATION

A demand-driven broadcast framework, termed *broadcast on demand*, (BoD) is proposed in [52], that satisfies the temporal constraints of the requests and uses scheduling techniques at the server side to dynamically utilize the limited bandwidth. The framework allows mobile clients limited ability to transmit queries to the server with the maximum tolerable latency. The server is capable of producing a broadcast that satisfies the clients' requests and retains scalability and bandwidth utilization. Essentially the broadcast communication is combined with on-demand data dissemination. It customizes broadcasts to service individual clients better, while avoiding the scalability problem of the client/server model. Time division multiplexing is used to utilize a fraction of the bandwidth for periodic broadcasts and the rest for on-demand data transmission.

    The broadcast strategy uses *earliest deadline first* (EDF) to schedule the transmission of data items. In the planning-based nonpreemptive broadcast strategy, a sorted target set of the number of requests to be broadcast is formed. At every iteration, an item having the closest deadline is chosen from the target. However, this schedule often performs poorly in overload situations. This is solved by using batching of multiple information and handling the batched requests by a single transmission of data items. For every transmission request, EDF-BATCH checks if that transmission is already planned. If so, it does not retransmit the data because the planned transmission will take care of that data; otherwise, the scheduler attempts to transmit the data. This results in bandwidth savings with less overhead. This strategy is extended to make the scheduling hybrid by incorporating on-demand pull-based scheduling schemes. When a client's request arrives the server first checks if a periodic broadcast can satisfy the request within deadline. If so, no on-demand scheduling is needed; otherwise the on-demand scheduling is used.

2.4.3 Channel Allocation for Data Dissemination

A different dynamic channel allocation method that assigns channels for broadcast or on-demand services based on system workload is discussed in [20, 27, 28]. The proposed channel allocation algorithm efficiently achieves the optimal channel allocation by approximation techniques. The wireless communication platform is assumed to be supported by a mobile support station (MSS). Every cell is assumed to consist of one MSS and multiple mobile computers. The MSS maintains D data items and the mobile computers issue requests to the MSS. Using the concept of a $M/M/c/n$ queuing model (with finite buffers) the expected access time ($E[PULL]$) of on-demand system under heavy load is approximated. Similarly for broadcast channels a cost analysis is performed and the expected access time for retrieving data through monitoring the broadcast channel is obtained. To achieve optimal data access efficiency in the cells, the system dynamically reassigns channels between on-demand and broadcast services. The allocation algorithm starts with exclusive on-demand system (i.e., the broadcast set being empty). It then initializes the lowest access time depending on whether the system is heavily or lightly loaded. Now at every iteration the algorithm identifies the data items to be transmitted. Then it computes the channel allocation policies and obtains the optimal allocation by choosing the policy that minimizes the overall access time. This scheme is performed both in heavy and light loads.

2.4.4 Wireless Hierarchical Data Dissemination System

A hierarchical data delivery (HDD) model is proposed in [21] that integrates data caching, information broadcasting, and point-to-point data delivery schemes. The broadcast schedule and cache management schemes are dynamically adjusted to minimize overall access time. Efficient data indexing methods are also explored in this environment. Data is stored in the hierarchies, with the most requested data in client cache, followed by commonly used data in the broadcast channel, and the least popular data in the server (for pulling). When a user issues a query the item is first searched for in the cache. It is retrieved if found in the cache; otherwise, the item is searched for in the server. If it is found within the threshold of the server's broadcast channel, it is obtained and kept in the cache; otherwise, it is explicitly pulled from the server. The clients can explicitly issue a signature to the broadcast channels. The model is formed using a single server and multiple clients. The client model is used to generate a query with Zipf's distribution and Gaussian distribution, broadcast channel monitoring, and request for pull items. The server model uses broadcast disk management techniques to schedule data items in an efficient manner.

2.4.5 ADAPTIVE HYBRID DATA DELIVERY

An adaptive hybrid data delivery strategy is also proposed in [32], that dynamically determines the popularity of the data items and effectively combines the push- and pull-based strategies. In other words, the data items are neither characterized nor predicted apriori. The strategy continuously adjusts the amount of bandwidth to match the diverse demand patterns of the clients. The total bandwidth is logically distributed into three parts for (1) broadcasting index block, (2) broadcasting data blocks, and (3) unicasting on-demand data blocks. The distribution adapts with the changes in clients' demands. The system begins with the server broadcasting one or more index or data objects. An increasing number of requests for a particular data will increase the bandwidth allocation for that data item and vice-versa. One major advantage of this approach is, the requests for a data item are reduced due to the satisfaction of the clients who recently received that data item. The server then reduces the bandwidth allocation for that data item. However, subsequent requests by the set of clients for that same data item increases the popularity of that item, and the server reassigns more bandwidth for that particular data item. One prime objective is to minimize the overall access time, in which the access time is composed of access time for index, tuning time, and access time for data.

2.4.6 ADAPTIVE REALTIME BANDWIDTH ALLOCATION

The real-time data delivery strategy discussed in [30] maintains a certain level of on-demand request arrival rate to get a close approximation of optimal system response time. One advantage of the system is that it does not explicitly need to know the access information of the data items. A single broadcast channel and a set of on-demand point-to-point channels are used in a single cell environment. The data items are of uniform size and the number of data items in the broadcast channel changes with variation in the system load. The clients first listen to the broadcast channels for respective data items they are waiting for. Only if the required data item is not found, the client transmits an explicit request to the server for that particular data item. An MFA (bit) vector and a broadcast number is kept. Each bit in the vector represents a data item in the broadcast channel. Whenever a request is satisfied, the corresponding bit in the vector is set. The server maintains a broadcast version number to ensure the validity of the relationship between bit-positions and data items. This vector and broadcast version number is piggy-backed to the server along with the on-demand data request. The server uses this to update the request information available.

2.4.7   ADAPTIVE DISSEMINATION IN TIME-CRITICAL ENVIRONMENTS

Adaptive, online, hybrid scheduling and data transmission schemes for minimizing the number of deadlines missed are also proposed in [16]. The information server dynamically adapts to the specific data items that need to be periodically broadcast and the amount of bandwidth assigned to each transmission mode. A *time critical adaptive hybrid broadcast* (TC-AHB) is proposed that combines both periodic broadcast and on-demand dissemination efficiently. In this scheme both the data items being broadcast and the amount of bandwidth assigned dynamically change in a per-cycle basis to adapt to clients' needs. The decision regarding periodic broadcast and on-demand transmission is dependent on the access frequency. The amount of bandwidth assigned, on the other hand, is related to the deadline constraints. The server always computes a periodic broadcast program for the next cycle and leaves some bandwidth for on-demand transmission. The broadcast frequency is the minimum needed to satisfy the deadline constraints of the specific data items. An online scheduling policy is used to prioritize the requests according to their deadlines and subsequently minimize the number of deadlines missed. The server broadcasts the items with high access requests and low bandwidth requirement. In each broadcast cycle the server includes the data item that aids in maximum bandwidth savings. This process is continued until some bandwidth is left for on-demand transmission. Such a greedy strategy is a near optimal solution. The on-demand scheduling used *earliest deadline first* (EDF), which is implemented using priority queues where priorities are inversely proportional to deadlines.

2.4.8   ADAPTIVE SCHEDULING WITH LOAN-BASED FEEDBACK CONTROL

To solve the dynamic information transmission problem, the work in [22] proposed a strategy to subsume the dynamic and static information into groups and introduce a loan-based slot allocation and feedback control scheme to effectively allocate the required bandwidth. A uniform building block, termed group, is designed. Every block has a unique Group-Id (GID). Two types of groups, namely, virtual and actual groups are formed. Clients interested in a static data item forms the virtual group. The server broadcasts the static items to the group at the same time. On the other hand, the actual group consists of the clients requesting dynamic data items. The server allocates a specific number of slots to each group depending on the particular group-popularity.

The dynamics of traffic might lead to an excess of scarce slots (bandwidth) to the groups. A loan-based slot allocation and feedback control (LSAF) scheme is introduced to complement the GID mechanism. At the start of a broadcast cycle, every group is assigned a slot-quota. The server then performs dynamic slot allocation among the groups during a broadcast cycle. When the slot-quota of a particular group is exhausted (due to transmission of different data items belonging to that group), the server attempts to loan a slot from another group to broadcast any data item belonging to the previous group. This loan for slots is determined by any one of three schemes: (1) sensitive loan: the server estimates and predicts the slot requirements of every group and loans a slot from the group, which will be having largest remaining slots in future; (2) insensitive loan: the server loans the slot from the group currently having the largest number of unused slots normalized by the slot quota; (3) greedy loan: the server takes the loan from the group having the largest remaining number of slots at the current instant. At the end of each broadcast cycle, the server estimates and learns the amount of slots taken to deliver all group-specific items of any group by a direct feedback mechanism. This feedback mechanism essentially gives the required slot-quota to meet the specific group's need. This also gives an estimate of dynamic item production and transmission rate. In order to meet the real-time constraints, the server also delivers the queued items using a pull-scheduling and has the capability of preempting the slots in the next broadcast cycle and broadcasts the queued items using a push scheduling.

2.4.9 FRAMEWORK FOR SCALABLE DISSEMINATION-BASED SYSTEMS

A general framework for describing and constructing dissemination-based information systems (DBIS) is described in [14]. By combining various data delivery techniques, the most efficient use available server, and communication resources, the scalability and performance of dissemination-oriented applications are enhanced.

The approach distinguishes between three types of nodes: (1) data sources provide base data for application, (2) clients consume this information, and (3) information brokers add value to information and redistributes it. Information brokers bind the different modes of data delivery and drive the scheduling to select a particular mode, depending on its access patterns. Brokers provide network transparency to the clients. Brokers can also be the data sources. Data can be cached at any of the many points along the data path from the server to the client. Cache invalidations and refresh messages need to be sent to each client cache manager. LRU or some other cache replacement policy can be used in this approach. Intermediate nodes can simply pass/propagate the data or can also perform some computations over those data. Provisions are also kept to recover some nodes from failure. The server relies on the clients' profile to optimize the push schedule. The framework provides techniques for delivering data in wide-area network settings in which nodes and links reflect extreme variation in their operating parameters. By adjusting the data delivery mechanism to match these characteristics, high performance and scalability can be achieved. The toolkit provides a set of classes that allow distributed nodes to negotiate and establish a connection and local cache. The data transmission policies need to be agreed upon between the server and the clients.

2.4.10 GUARANTEED CONSISTENCY AND CURRENCY IN READ-ONLY DATA

To ensure various degrees of data consistency and currency for read-only transactions, various new isolation levels are proposed in [44]. Efficient implementation of these isolation levels is also proposed. This is used in a hybrid data transmission environment. The newly proposed consistency levels are independent of the existing concurrency protocols.

Although serializability is standard criteria for transaction processing in both stationary and mobile computing, it is, in itself, not sufficient for preventing read-only transactions from experiencing anomalies related to data currency. A start-time multi-version serialization graph (ST-MVSG) is a directed graph with nodes = *commit* (*MVH*) and edges E such that there is an edge representing every arbitrary dependency. Let MVH be a multiversion history over a set of committed transactions. Then MVH is BOT serializable if ST-MVSG is acyclic. In an MVH that contains a set of read-write transactions such that all read-write transactions are serializable, each read-only transaction satisfying READ-RULE is also serializable. This leads to the conclusion that MVH is strict forward BOT serializable if SFR-MVSG is serializable. In a multiversion history containing a set of read-write transactions such that all read-write transactions are serializable, each read-only transaction is serializable with respect to transactions belonging to the corresponding update. Simulation results demonstrate that this improves the throughput control and number of abort associated in transactions.

2.4.11 BROADCAST IN WIRELESS NETWORKS WITH USER RETRIALS

Most of the research works in data broadcast do not consider the possibility of a single user making multiple-request submission attempts. Such retrial phenomenon has significant affect on the system's overall performance. The prime objective of the work in [50] is to capture and analyze the user retrial phenomenon in wireless data broadcast schemes. The objective is realized by introducing different performance measures, like broadcast and unicast service ratio, service loss, waiting time, and reneging probability. Based on the analytical expressions for these performance measures, the existence of a single, optimal broadcast scheduling scheme is proved. The solution provides optimal performance with respect to system's throughput, grade, and quality of service. This method is extended to design a hybrid unicast/broadcast scheduling scheme with user's retrials.

2.5   SUMMARY

Basically all scheduling can be divided into push and pull scheduling schemes. However, both scheduling schemes have their own limitations. Hence, a suitable combination of push and pull schemes is required to develop a hybrid scheduling strategy that has the capability of improving the overall performance. In this chapter we presented a broad overview of the major existing push, pull, and hybrid scheduling strategies. While most of the strategies attempt to minimize the client's waiting time, some are also focused on delay jitter, overall cost, and consistency.

3 Hybrid Push-Pull Scheduling

In this chapter we introduce a new hybrid push-pull scheduling strategy. In short, the strategy partitions the entire set of items into two disjoint sets: the access set and the request set that are the push and pull subsets, respectively. The access set contains the hot data items that are the items that are very popular among clients and hence would be sent based on some push scheduling algorithm. The request set contains cold data items, which are not so popular items, and hence would be sent on-demand based on some pull scheduling algorithm as shown in Figure 3.1. The schedule strictly alternates between a push and a pull operation to transmit all the data items. While initially the system operates on unit-length, homogeneous data items, the work is extended to also include the heterogeneous, variable-length items. The selection criteria for a push item is based on packet fair scheduling and a pull item is selected on the basis of most requested first (MRF) (for homogeneous items) and stretch-optimal scheduling (for heterogeneous items). The scheme is further enhanced to incorporate the role of client priorities to resolve the tie. Suitable performance modelling is done to analyze the average system performance. Simulation experiments support this performance analysis and point out the efficiency of the hybrid system in reducing the overall average access time.

3.1 HYBRID SCHEDULING FOR UNIT-LENGTH ITEMS

Before introducing our proposed hybrid scheduling for unit-length data items, we first highlight the assumptions we have used in our hybrid scheduling system.

3.1.1 ASSUMPTIONS AND MOTIVATIONS

1. We assume a system with a single server and multiple clients thereby imposing an asymmetry. Figure 3.2 shows the schematic diagram of such an asymmetric environment consisting of a single server and multiple clients. The uplink bandwidth is much less than the downlink bandwidth.

2. The database at the server is assumed to be composed of D total number of distinct data items, each of unit length.

3. The access probability P_i of item i is a measure of its degree of popularity. In order to keep a wide range of access probabilities (from very skewed to very similar), we assume that the access probabilities P_i follow the Zipf's distribution with access *skew-coefficient* θ: $P_i = \frac{(1/i)^\theta}{\sum_{j=1}^{n}(1/j)^\theta}$. It is assumed that the server knows the access probability of each item in advance. The items are numbered from 1 to D in decreasing order of their access probability, thus $P_1 \geq P_2 \geq \ldots \geq P_D$. Clearly, from time to time, the server recomputes the access probability of the items, renumbers them

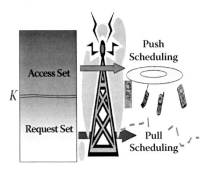

FIGURE 3.1 Our hybrid scheduling strategy.

as necessary, and eventually makes available to all clients the new numbering of the items. It is assumed that one unit of time is the time required to spread an item of unit length.

Our hybrid scheduling algorithm for the client side is depicted in Figure 3.3. As shown in the figure, the clients simply send the request for an item to the server (irrespective of the fact whether the item is in the push set or the pull set) and then waits and listens to the downlink channel to grab the data of its interest, whenever it goes by. We say that the client accesses an item if that item is pushed, while that item is requested if the item is pulled. Moreover, let the load N of the system be the number of requests/access arriving in the system for unit of time. Let the access time, $T_{acc,I}$ be the amount of time that a client waits for a data item i to be broadcast after it begins to listen. Moreover, let the response time, $T_{res,I}$ be the amount of time between the request of item i by the client and the data transmission. Clearly, the aim of push scheduling is to keep the access time for each push item i as small as possible, while that of pull scheduling is to minimize the response time for each pull item i. In a push-based system, one of the overall measures of the scheduling performance is called average expected access time, $T_{exp-acc}$, which is defined as $T_{exp-acc} = \sum_{i=1}^{D} P_i \cdot \overline{T_{acc,i}}$, where $\overline{T_{acc,i}}$ is the average expected access time for item i. If instances are equally spaced in the broadcast cycle, then $\overline{T_{acc,i}} = \frac{s_i}{2}$, where s_i is the spacing between the two instances of same item i. The push scheduling is

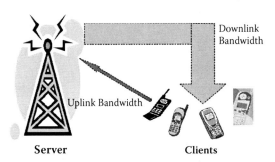

FIGURE 3.2 Concept of asymmetric environment.

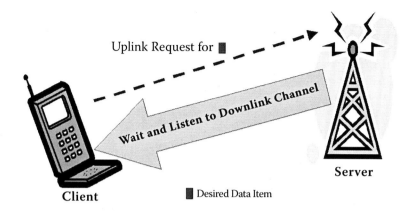

FIGURE 3.3 Our hybrid scheduling for the client side.

based on the packet fair scheduling algorithm 2.1.4. Similarly for the pull scheduling, it can be defined the average expected response time, denoted $T_{exp-res}$.

To explain the rationale behind our approach, let us first describe in detail the intuition behind the hybrid scheduling in [18] and point out some of its drawbacks. To make the average expected access time of the system smaller, the solution in [18] flushes the pull queue. Figure 3.4, shows an example of the broadcast schedule for [18]. The figure shows that the server pushes one item followed by three pull items, assuming that the pull queue has three items in it at that moment. Again the system pushes one item followed by one pull item, if only one pull item was present in the system's pull queue. In the next part of the broadcast cycle, the server pushes two consecutive, items assuming that the pull queue was empty, leaving no items to be pulled as shown in Figure 3.4.

Let the *push set* consist of the data items numbered from 1 up to K, termed from now on the *cutoff point*, and let the remaining items from $K+1$ up to D form the *pull set*. Hence, the average expected waiting time for the hybrid scheduling is defined as:

$$T_{exp-hyb} = T_{exp-acc} + T_{exp-res} = \sum_{i=1}^{K} P_i \cdot \overline{T_{acc,i}} + \sum_{i=K+1}^{D} P_i \cdot \overline{T_{res,i}}$$

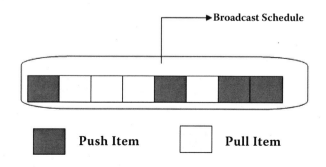

FIGURE 3.4 Push one pull all schedule.

As the push-set becomes smaller, the average expected access time $T_{exp-acc}$ becomes shorter. However, the pull-set size becomes larger, leading to a longer expected response time $T_{exp-res}$. The size of the pull-set might also increase the average access time $\overline{T_{acc,i}}$, for every push item. In fact, if the hybrid scheduling serves, between any two items of the cyclic push scheduling, all the pending requests for pull items in first-come-first-served (FCFS) order, it holds for the average expected access time for item i : $\overline{T_{acc,i}} = (s_i + s_i \cdot q)/2$, where q is the average number of distinct pull items for which arrives at least one pending request in the pull queue for unit of time. From now on, we refer to q as the dilation factor of the push scheduling. To limit the growth of the $\overline{T_{acc,i}}$, and therefore that of the $\overline{T_{exp-acc}}$, the push set taken in [18] is large enough that, on average, no more than one request for all the pull items arrives during a single unit time. To guarantee a dilation factor q equal to 1 when the system load is equal to N, [18] introduces the concept of the build-up point B. B is the minimum index between 1 and D for which it holds $N(1 - \sum_{i=1}^{B} P_i) \leq 1$, where N is the average access/requests arriving at unit of time. In other words, [18] pushes all the items from 1 up to B to guarantee that no more than one item is waiting in the pull queue to be disseminated, and therefore to achieve $q = 1$. Thus, through the build-up point, the dilation factor q is guaranteed to be equal to 1, hence making the broadcast schedule of Figure 3.4 look like Figure 3.5, thereby guaranteeing one push followed by one pull item.

After having bounded the dilation factor to 1, [18] chooses as the cutoff point between the push and pull items the value K, with $K \geq B$, such that K minimizes the *average* expected waiting time for the hybrid system. Intuitively, the partition between push and pull items found out in [18] is meaningful only when the system load N is small and the access probabilities are much skewed. Under these conditions, indeed, the build-up point B is low. Hence, there may be a cutoff K, such that $B \leq K \leq D$, which improves on the average expected access time of the pure-push system. However, when either the system has a high load N or all items have almost the same degree of probability, the distinction between the high and low demand items becomes vague and artificial; hence, the value of build-up point B increases, finally leading to the maximum number D of items in the system. In those cases, the solution proposed in [18] almost always behaves as a pure push-based system.

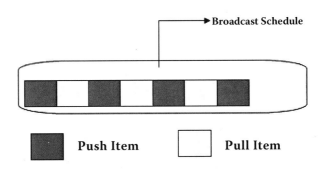

FIGURE 3.5 Push one pull one schedule.

3.1.2 THE BASIC HYBRID PUSH-PULL ALGORITHM

We now present a hybrid scheduling scheme that improves on [18] when the load is high or when the access probabilities are balanced, that is, when the scheduling in [18] reduces to the pure-push scheduling. The solution proposed in this chapter again partitions the data items in the push set and the pull set, but it chooses the value of the cutoff point K between those two sets independent of the buildup point. Indeed, we let the pull queue grow in size, and the push set can contain any number of data items. After each single broadcast, we do not flush out the pull queue, which may contain several different pending requests. In contrast, we just pull one single item: the item, that has the largest number of pending requests. We term this pull scheduling as max-request-first (MRF) scheduling, that is, we choose the item that has gathered the maximum number of requests or that has the maximum number of clients waiting, first to pull from the pull queue. In case of a tie, FCFS policy is applied. The MRF policy is shown in Figure 3.6.

Note that simultaneously with every push and pull, N more access/requests arrive to the server, thus the pull queue grows drastically at the beginning. In particular, if the pull set consists of the items from $K+1$ up to D, at most $N \times \sum_{j=K+1}^{D} P_i$ requests can be inserted in the pull queue at every instance of time, out of which, only one, the pull item that has accumulated the largest number of requests, is extracted from the queue to be pulled. We are sure, however, that the number of distinct items in the pull queue cannot grow uncontrolled because the pull queue can store, at most, only as many distinct items as are in the pull set; that is, no more than $D - K$ items. So, eventually the new arriving requests will only increase the number of clients waiting in the queue for some item, leaving unchanged the queue length. From this moment, we say that the system has reached a *steady state*. In other words, the pending requests will start to accumulate behind each pull item without further increasing the queue length. Hence, just pulling the high demanded pull item, the system will not serve just one client but many. A pull item cannot be stuck in the pull queue for more than as many units of time as the length of the queue. The push system, on the other hand, incurs an average delay of $\sum_{i=1}^{K} s_i P_i$, where $s_i = \dfrac{\sum_{j=1}^{k} \sqrt{\hat{P}_j}}{\sqrt{\hat{P}_i}}$, and $\hat{P}_i = \dfrac{P_i}{\sum_{j=1}^{k} P_j}$.

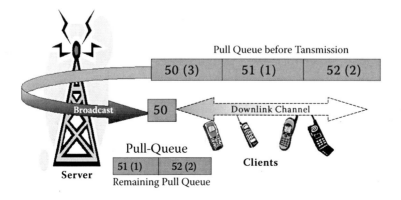

FIGURE 3.6 MRF policy for pull items.

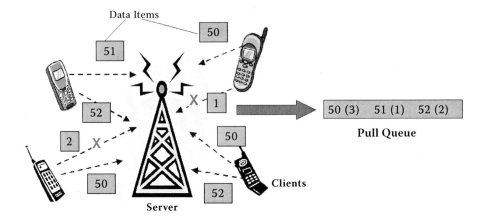

FIGURE 3.7 Formation of pull queue.

The schematic diagram of our basic hybrid scheduling algorithm is shown in Figure 3.7. The server performs several actions simultaneously. From one side, it monitors the access probabilities of the data items and the system load. When those parameters diverge significantly from the assumptions previously made by the system, the server renumbers the data items, and recalculates the cutoff point K to separate the push set from the pull set, as illustrated in Figure 3.8. Note that K is selected in such a way that the average expected waiting time of the hybrid scheduling $T_{exp-hyb}$ is minimized. In addition, the server listens to all the requests of the clients and manages the pull queue. The server ignores the requests that come for an item which is in the push set, as that item will be pushed (broadcast) by the server based on PFS, leaving the server to maintain only the pull queue.

Integer function CUTOFF POINT $(D, P = P_1, P_2 \dots P_D) : K$

/* D: Total No. of items in the Database of the server
P: Sorted vector of access probability of items in decreasing order
K: Optimal Cutoff Point */
$K := 1; T_{exp-hyb}(0) := T_{exp-hyb}(1) := D;$
while $K \leq D$ **and** $T_{exp-hyb}(K - 1) \geq T_{exp-hyb}(K)$ **do**
begin
Set $s_i = \dfrac{\sum_{j=1}^{k} \sqrt{\hat{P}_j}}{\sqrt{\hat{P}_i}}$, **where** $\hat{P}_i = \dfrac{P_i}{\sum_{j=1}^{K} P_j}$,
$T_{exp-hyb}(K) = \sum_{i=1}^{K} S_i P_i + \sum_{i=K+1}^{D} P_i * (D - K); K := K + 1;$
end
return $(K - 1)$

FIGURE 3.8 Algorithm to set the optimal cutoff point K.

```
Procedure HYBRID SCHEDULING;

while true do
begin
compute an item from the push scheduling;
broadcast that item;
if the pull queue is not empty, then
extract the most requested item from the pull queue,
clear the number of pending requests for that item;
Pull that item
end;
```

FIGURE 3.9 Algorithm at the server end.

The pull queue, implemented by a max-heap, keeps in its root, at any instant, the item with the highest number of pending requests. For any request i, if i is larger than the current cutoff point K, $i \geq K$, i is inserted into the pull queue, the number of the pending requests for i is increased by one, and the heap information updates accordingly. On the other hand, if i is smaller than or equal to K, $i \leq K$, the server simply drops the request because that item will be broadcast by the push scheduling. Finally, the server is in charge of deciding at each instant of time which item must be spread. The scheduling is derived as explained in Figure 3.9, where the details for obtaining the push scheduling (PFS) are omitted. Interested readers can find it in Hameed and Viadya (1999). To retrieve a data item, a client performs the actions mentioned in Figure 3.10.

3.2 SIMULATION EXPERIMENTS

First of all, we compare the simulation results of the new algorithm with those of the hybrid scheduling in [18], with the results of the pure-push scheduling, and with the analytic expression used to derive the optimal cutoff point. We run experiments for $D = 100$, for the total number of access or requests in the system $M = 25000$ and for $N = 10$ or $N = 20$. The results are reported in Tables 3.1 and 3.2, respectively for

```
Procedure CLIENT-REQUEST (i):

/* i : the item the client is interested in */
begin
send to the server the request for item i;
wait until get i from the channel
end
```

FIGURE 3.10 Algorithm at the client side.

TABLE 3.1
Expected Access Time for Several Values of θ and $N = 10$

θ	New Hybrid	Old Hybrid [18]	Push	Analytical
0.50	40.30	44.30	45.03	36.01
0.60	37.78	42.35	43.01	36.21
0.70	35.23	40.01	40.50	35.04
0.80	32.36	37.31	37.47	33.56
0.90	29.38	34.12	34.30	29.94
1.00	25.95	29.93	30.90	29.73
1.10	22.95	24.38	27.75	27.09
1.20	19.90	20.61	24.50	25.19
1.30	17.04	17.04	20.86	22.51

$N = 10$ and $N = 20$. For both Tables 3.1 and 3.2, the value of θ is varied from 0.50 to 1.30, to have the access probabilities of the items initially from similar to very skewed. Note that for θ no larger than 1, the analytic average expected access time is close to that measured with the experiments. This confirms that, when the access probabilities are similar, the pull items remain in the pull queue for a time no larger than the total number of pull items that is $D - K$. For larger values of θ, the experimental measure of the expected response time is smaller than that of the analytic expected value because the access probabilities are very skewed, fewer than $D - K$ items can be present simultaneously in the pull queue. Therefore, the actual waiting time of the client is eventually shorter than $D - K$. Further experimental results have shown that when θ is varied from 0.90 to 1.30, the length of the pull queue is approximated better by the value $D \times \sum_{i=K+1}^{D} P_i$ than by $D - K$. Moreover, as discussed earlier, when the system is highly loaded, the scheduling algorithm in [18], whose cutoff point K must be larger than the buildup point B, almost reduces to the pure-push scheduling. Contradictory to [18], in the new hybrid algorithm, even with a very high loaded system, results are better than in a pure push-based system,

TABLE 3.2
Expected Access Time for Several Values of θ and $N = 20$

θ	New Hybrid	Old Hybrid [18]	Push	Analytical
0.50	41.96	44.44	44.70	37.59
0.60	39.39	42.45	42.61	35.68
0.70	36.59	40.10	40.30	34.53
0.80	33.49	37.39	37.65	33.06
0.90	30.39	33.78	34.12	29.94
1.00	27.20	30.69	37.68	29.73
1.10	23.88	27.54	27.71	27.09
1.20	21.14	23.23	23.94	24.43
1.30	18.26	19.49	21.07	24.24

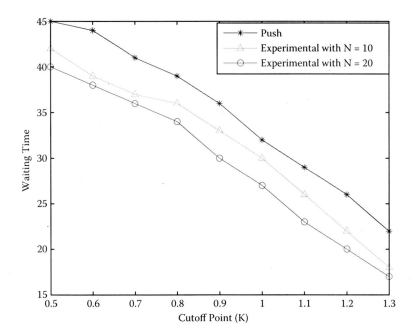

FIGURE 3.11 Comparison between push scheduling and new algorithm.

as illustrated in Figure 3.11. Besides, in Figure 3.12, the values of the cutoff point K for our solution, which takes K independent of B, and for the hybrid scheduling proposed in [18] are depicted for $N = 10$ and $N = 20$.

3.3 DYNAMIC HYBRID SCHEDULING WITH HETEROGENEOUS ITEMS

Although the above mentioned hybrid scheduling algorithm outperforms the existing hybrid scheduling algorithms a close look at it proves that it is not practical for real systems. The reason is it assumes that all the data items in the server to be of same (unit) length Figure 3.13, which is not the case practically.

A more practical case would be where the server has many items, each being a different length. Thus the previously mentioned hybrid scheduling algorithm is extended to incorporate the heterogeneous data items and to resolve a tie while selecting a pull item [35, 41]. This variation in length of the items results in differences in service time. For example, if we consider a scenario in which the pull queue has the same number of maximum requests for two data items A and B having lengths 10 and 2 units, respectively, if we broadcast A first followed by B, the average waiting time of the system is six units of time assuming items of unit length take one unit of time to be broadcast; whereas when the system broadcasts B followed by A, the average waiting time of the system could be reduced to two units. This is shown in Figure 3.14. Thus we argue that MRF alone could no longer be a fair scheduling for such a heterogenous system.

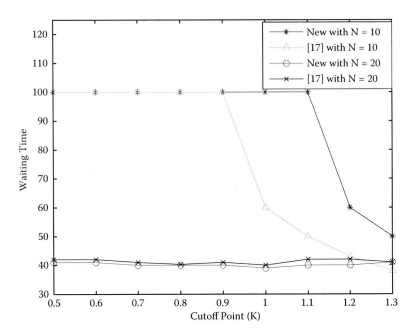

FIGURE 3.12 Cutoff point when $N = 10, 20$.

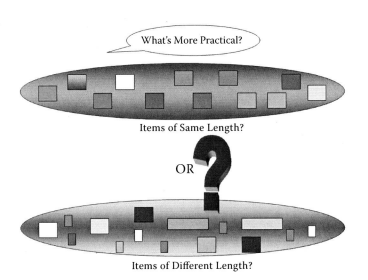

FIGURE 3.13 Length of the items.

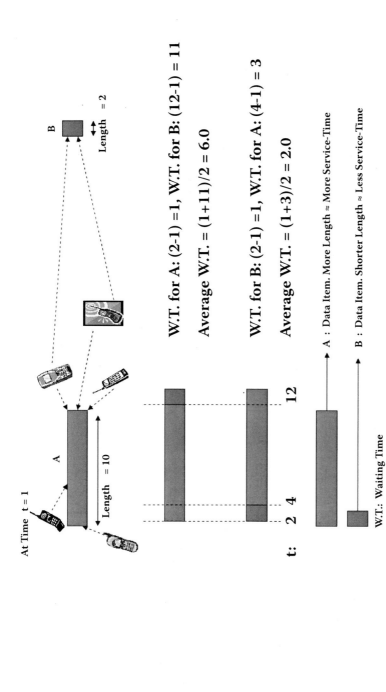

FIGURE 3.14 Effect of item length on average time of the system.

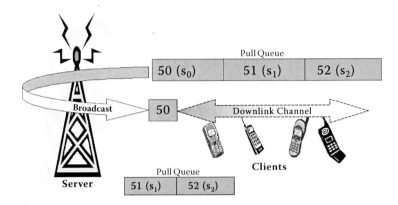

FIGURE 3.15 Stretch optimal pull scheduling.

Rather, the pull scheduling now needs to consider the item length along with the number of requests accumulated. This motivates us to use stretch-optimal scheduling which is based on the principle of maximum request minimum service time first policy.

3.3.1 HETEROGENEOUS HYBRID SCHEDULING ALGORITHM

We still assume an ideal environment with a single server serving multiple clients, thereby imposing asymmetry. As previously, the database at the server consists of a total number of D distinct items, out of which, K items are pushed and the remaining $(D - K)$ items are pulled. However, the items now have variable lengths, and each item i has a different access probability P_i. The service time for an item is dependent on the size of that item. The larger the length of an item, the higher its service time.

We have adopted PFS 2.1.4 in our hybrid algorithm as the push mechanism. As before, the term push scheduling will refer to the cyclic scheduling produced by the PFS algorithm applied to the push set. On the other hand, for the pull mechanism, we select the item that has maximum stretch value $\mathscr{S}_i = \frac{Request\ Count\ for\ item\ i}{Length_i^2}$. Such a stretch optimal pull scheduling is shown in Figure 3.15.

We have assumed an ideal environment, where the client needs to send the server its request for the required item i along with its unique ID and waits until it listens for i on the channel (see Figure 3.16).

Procedure CLIENT-REQUEST:

begin
```
    send to the server the request for a particular item
    with a unique id associated with the item;
    wait until listen for that item on the channel;
```
end

FIGURE 3.16 Client side algorithm.

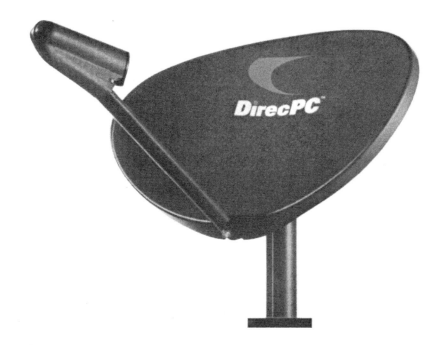

FIGURE 3.17 Hughes DirecPC dish.

Note that the behavior of the client is independent of the fact that the requested item belongs to the push set or the pull set. As previously mentioned in Section 1.3, the Huges Network Systems DirecPC architecture [23] is a suitable example for such a broadcast system. Figure 3.17 is the Hughes DirecPC dish.

The server maintains the database of all items. The system starts as a pure pull-based scheduler (i.e., the push set is empty) assuming that all the items have the same access probability and few requests occur. Then, based on the requests received for each item during a certain interval of time, it dynamically moves to a hybrid system with the data items separated into the push set and the pull set. Precisely, at regular intervals of time, the server monitors the data access probabilities of the items and the arrival rate of the requests. If the values registered by the server significantly deviate by the values for which the current segregation point between the push and the pull sets has been computed, the cutoff point must be recomputed.

Once the cutoff point is fixed, the server continuously alternates a push phase, in which a single item is broadcast, and a pull phase, in which a single item is disseminated, when there are clients waiting for pull items. After every push and every pull operation, the server accepts the set of requests arriving into the system. More precisely, the server simply collects statistics about the requests for the push items. After every push, if the pull queue is not empty, the server chooses the item with maximum stretch value. It might happen that more than one item has the same stretch value. In that case, the server considers the item that has maximum priority. Priorities of the items are estimated by adding the priorities of the clients requesting that particular item, and then normalizing it. The ID of the client is used by the server

```
Procedure Hybrid Scheduling;

while (true) do
begin
    Push-Phase:
        broadcast an item selected according to the packet fair scheduling;
        handle the requests occurring during the push-phase;
    if the pull queue is not empty then
    Pull-Phase:
        extract from the pull queue the item whose stretch is maximum;
        if tie
          extract the item whose sum of the clients' priority is high;
          if tie
            extract the item with the smallest index;
        clear the number of pending requests for this item, and pull it;
        handle the requests occurring during the pull-phase;
end;
```

FIGURE 3.18 Hybrid scheduling algorithm.

to calculate its priority. Figure 3.18 provides the pseudo-code of the hybrid scheduling algorithm executing at the server side while the push and pull sets do not change.

3.3.2 MODELING THE SYSTEM

In this section we evaluate the performance of our hybrid scheduling by developing suitable analytical models. The goal of this analysis is two-fold: it is used to estimate the minimum expected waiting time (delay) of the hybrid system when the size of the push set is known, and to determine the cutoff point (K) between the push set and pull set when the system conditions (arrival rate and access probabilities) change. Because the waiting time is dependent on the size K of the push set, we investigate, by the analytical model, into the delay dynamics for different values of K to derive the cutoff point; that is, the value of K that minimizes the system delay.

Before proceeding, let us enumerate the parameters and assumptions used in our model:

1. The database consists of $D = \{1, \ldots, D\}$ distinct items, sorted by non-increasing access probabilities $\{P_1 \geq \ldots \geq P_D\}$. Basically, the access probability gives a measure of an item's popularity among the clients. We have assumed that the access probabilities (P_i) follow the *Zipf's distribution* with *access skew coefficient* θ, such that $P_i = \frac{(1/i)^\theta}{\sum_{j=1}^n (1/j)^\theta}$. Every item has a different length randomly distributed between 1–L, where L is the maximum length.

2. Let C, K, and $\rho_{(cl)}$, respectively, denote the maximum number of clients, the size of the push set, and priority of client cl. The server pushes K items and clients pull the remaining $(D - K)$ items. Thus, the total probability

TABLE 3.3
Symbols Used for Performance Analysis

Symbols	Descriptions
D	Maximum number of items
C	Maximum number of clients
i	Index of data item
K	Size of the push set
P_i	Access probability of item i
L_i	Length of item i
λ	Pull queue arrival rate
$\lambda_{arrival}$	System arrival rate
μ_1	Push queue service rate
μ_2	Pull queue service rate
S_i	Space between the two instances of data item i
$\rho_{(cl)}$	Priority of client cl
ρ_i	Priority of data item i
$E[W_{pull}]$	Expected waiting time of pull system
$E[W^q_{pull}]$	Expected waiting time of pull queue
$E[\mathscr{L}_{pull}]$	Expected number of items in the pull system
$E[\mathscr{L}^q_{pull}]$	Expected number of items in the pull queue

of items in the push set and pull set are respectively given by $\sum_{i=1}^{K} P_i$ and $\sum_{i=K+1}^{D} P_i = (1 - \sum_{i=1}^{K} P_i)$.

3. The service times of both the push and pull systems are exponentially distributed with mean μ_1 and μ_2, respectively.

4. The arrival rate in the entire system is assumed to obey the Poisson distribution with mean $\lambda_{arrival}$.

Table 3.3 lists the symbols with their meanings used in the context of our analysis. Now, we are in a position to analyze the system performance for achieving the minimal waiting time.

3.3.2.1 Minimal Expected Waiting Time

Figure 3.19 illustrates the birth and death process of our system model, where the arrival rate in the pull system is given by $\lambda = (1 - \sum_{i=1}^{K} P_i)\lambda_{arrival}$. First, we discuss the state space and possible transitions in this model.

1. Any state of the overall system is represented by the tuple (i, j), where i represents the number of items in the pull system and $j = 0$ (or 1) respectively represents whether the push system (or pull system) is being served.

2. The arrival of a data item in the pull system results in the transition from state (i, j) to state $(i + 1, j)$, for $0 \leq i \leq C$ and $0 \leq j \leq 1$. However, the service of an item results in two different actions. Because the push system

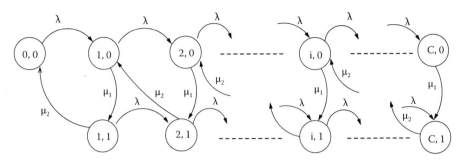

FIGURE 3.19 Performance modeling of our hybrid system.

is governed by packet fair scheduling, the service of an item in the push queue results in transition from state $(i,0)$ to state $(i,1)$, for $0 \leq i \leq C$. On the other hand, the service of an item in the pull queue results in transition from state $(i, 1)$ to the state $(i - 1,0)$, for $1 \leq i \leq C$.

3. Naturally, the state $(0,0)$ of the system represents that the pull-queue is empty and any subsequent service of the items in the push system leaves it in the same $(0,0)$ state. Obviously, state $(0,1)$ is not valid because the service of an empty pull-queue is not possible.

In the steady state, using the flow balance conditions of Chapman–Kolmogrov's equation [17], we have the following equation for representing the initial system behavior:

$$p(0, 0) \lambda = p(1, 1) \mu_2 \tag{3.1}$$

where $p(i, j)$ represents the probability of state (i, j). The overall behavior of the system for push (upper chain in Figure 3.19) and the pull system (lower chain in Figure 3.19) is given by the following two generalized equations:

$$p(i, 0)(\lambda + \mu_1) = p(i - 1, 0)\lambda + p(i + 1, 1)\mu_2 \tag{3.2}$$

$$p(i, 1)(\lambda + \mu_2) = p(i, 0)\mu_1 + p(i - 1, 1)\lambda \tag{3.3}$$

The most efficient way to solve equations (3.2) and (3.3) is using the z-transforms [17]. The resulting solutions are of the form:

$$P_1(z) = \sum_{i=0}^{C} p(i, 0) z^i \tag{3.4}$$

$$P_2(z) = \sum_{i=0}^{C} p(i, 1) z^i \tag{3.5}$$

Now, dividing both sides of equation (3.2) by μ_2, letting $\rho = \frac{\lambda}{\mu_2}$ and $f = \frac{\mu_1}{\mu_2}$, performing subsequent z-transform as in equation (3.4) and using equation (3.1), we obtain

$$P_2(z) = p(1, 1) + z(\rho + f)[P_1(z) - p(0, 0)] - \rho z^2 P_1(z)$$
$$= \rho \, p(0, 0) + z(\rho + f)[P_1(z) - p(0, 0)] - \rho z^2 P_1(z) \tag{3.6}$$

Similarly, transforming equation (3.3) and performing subsequent derivations, we get

$$P_2(z) = \frac{f\,[P_1(z) - p(0,0)]}{(1 + \rho - \rho\,z)} \tag{3.7}$$

Now, estimating the system behavior at the initial condition, we state the following normalization criteria:

1. The occupancy of pull states is the total traffic of pull queue and given by:
   $P_2(1) = \sum_{i=1}^{C} p(i,1) = \rho$.
2. The occupancy of the push states (upper chain) is similarly given by:
   $P_1(1) = \sum_{i=1}^{C} p(i,0) = (1 - \rho)$.

Using these two relations in Equation (3.6), the idle probability, $p(0,0)$, is obtained as follows:

$$P_2(1) = \rho\,p(0,0) + (\rho + f)\,[P_1(1) - p(0,0)] - \rho\,P_1(1)$$
$$\rho = \rho\,p(0,0) + (\rho + f)\,[1 - \rho - p(0,0)] - \rho\,(1 - \rho)$$
$$= f(1 - \rho) - f\,p(0,0)$$
$$f\,p(0,0) = f\,(1 - \rho) - \rho$$
$$p(0,0) = 1 - \rho - \frac{\rho}{f}$$
$$= 1 - 2\,\rho, \text{ (if } \mu_1 = \mu_2) \tag{3.8}$$

Generalized solutions of equations (3.6) and (3.7) to obtain all values of probabilities $p(i, j)$ become very complicated. Thus, the best possible way is to go for an expected measure of system performance, such as the average number of elements in the system and average waiting time. The most convenient way to derive the expected system performance is to differentiate the z-transformed variables, $P_1(z)$ and $P_2(z)$ and capture their values at $z = 1$. Thus, differentiating both sides of equation (3.6) with respect to z at $z = 1$, we estimate the expected number of items in the pull system, $E[\mathscr{L}_{pull}]$, as follows:

$$\left[\frac{d P_2(z)}{dz}\right]_{z=1} = (\rho + f)\left[\frac{d P_1(z)}{dz}\right]_{z=1} + P_1(1)\,(f - \rho)$$
$$- p(0,0)\,(\rho + f) - \rho\left[\frac{d P_1(z)}{dz}\right]_{z=1}$$

$$E[\mathscr{L}_{pull}] = (\rho + f)\,\mathscr{N} + (1 - \rho) - (\rho + f)\left(1 - \rho - \frac{\rho}{f}\right) - \rho\,\mathscr{N}$$
$$= \left(\frac{\mu_1}{\mu_2}\right)\mathscr{N} + \left(1 - \frac{\lambda}{\mu_2}\right) - \left(\frac{\lambda + \mu_1}{\mu_2}\right)\left(1 - \frac{\lambda}{\mu_2} - \frac{\lambda}{\mu_2}\right)$$
$$= \mathscr{N} + \left(1 - \frac{\lambda}{\mu}\right) - \left(1 + \frac{\lambda}{\mu}\right)\left(1 - 2\frac{\lambda}{\mu}\right), \quad \text{(if } \mu_1 = \mu_2 = \mu) \tag{3.9}$$

where $\mathscr{N}$ is the average number of users waiting in the pull queue when push is being served. Once we have the expected number of items in the pull system from equation (3.9), using Little's formula [17] we can easily estimate the average waiting time of the system, $E[W_{pull}]$, average waiting time of the pull queue, $E[W^q_{pull}]$, and expected number of items, $E[\mathscr{L}^q_{pull}]$, in the pull queue as follows:

$$E[W_{pull}] = \frac{E[\mathscr{L}_{pull}]}{\lambda}$$

$$E[\mathscr{L}^q_{pull}] = E[\mathscr{L}_{pull}] - \frac{\lambda}{\mu_2}$$

$$E[W^q_{pull}] = E[W_{pull}] - \frac{1}{\mu_2} \tag{3.10}$$

Note that there is a subtle difference between the concept of pull system and pull queue. While the pull queue considers only the items waiting for service in the queue, the pull system also includes the item(s) currently being serviced. However, the expected waiting time for the pull system discussed previously does not consider the priorities associated with the individual data items. Such estimates can fulfill the need for average system performance when every item in the pull queue has accumulated a different number of requests. However, when any two items contain the same number of pending requests, the priorities associated with those two items come into consideration. This will affect the arrival and service of the individual data items. Thus, a smart system should consider the priorities of the data items influenced by the client priorities.

3.3.2.1.1 Role of Client Priorities

Any client j is associated with a certain priority $\rho_{(j)}$ that reveals its importance. The priority of a particular data item is derived from the total normalized priorities of all the clients currently requesting that data item. Thus, if a particular data item i is requested by a set $\mathscr{C}$ of clients, then its priority is estimated as: $\rho_i = \frac{1}{|\mathscr{C}|} \times \sum_{\forall j \in \mathscr{C}} \rho_{(j)}$. The lower the value of $\rho_{(cl)}$, the higher the priority. When two items have the same stretch value, the item with higher priority is serviced first. This is also practical since such an item is requested by more important clients than its counterpart. Considering a nonpreemptive system with many priorities, let us assume the data items with priority ρ_i have an arrival rate λ_i and service time μ_{2_i}. The occupancy arising due to this jth data item is represented by $\rho_i = \frac{\lambda_i}{\mu_{2_i}}$, for $1 \leq i \leq$ max, where max represents maximum possible value of priority. Also, let $\sigma_i = \sum_{x=1}^{i} \rho_x$. In the boundary conditions, we have $\sigma_0 = 0$ and $\sigma_{\max} = \rho$. If we assume that a data item of priority x arrives at time t_0 and gets serviced at time t_1, then the waiting time is $t_1 - t_0$.

Let at t_0, there be n_i data items present having priorities i. Also, let S_0 be the time required to finish the data item already in service, and S_i be the total time required to serve n_i. During the waiting time of any data item, n'_i new items having the same number of pending requests and higher priority can arrive and go to service before the current item. If S'_i is the total time required to service all the n'_i items, then the expected waiting time will be

$$E\left[W^{q(x)}_{pull}\right] = \sum_{i=1}^{x-1} E[S'_i] + \sum_{i=1}^{x} E[S_i] + E[S_0] \tag{3.11}$$

To get a reasonable estimate of $W_{pull}^{q(i)}$, three components of equation (3.11) need to be evaluated individually.

(i) *Estimating* $E[S_0]$: The random variable S_0 actually represents the remaining service time, and achieves a value 0 for idle system. Thus, the computation of $E[S_0]$ is performed in the following way:

$$E[S_0] = Pr[\text{Busy System}] \times E[S_0|\text{Busy System}]$$

$$= \rho \sum_{i=1}^{max} E[S_0|\text{Serving items having priority i}]$$

$$\times Pr[\text{items having priority i}]$$

$$= \rho \sum_{i=1}^{max} \frac{\rho_i}{\rho \mu_{2_i}}$$

$$= \sum_{i=1}^{max} \frac{\rho_i}{\mu_{2_i}} \tag{3.12}$$

(ii) *Estimating* $E[S_i]$: The inherent independence of Poisson process gives the flexibility to assume the service time $S_i^{(n)}$ of all n_i customers to be independent. Thus, $E[S_i]$ can be estimated using the following steps:

$$E[S_i] = E\left[n_i S_i^{(n)}\right] = E[n_i]E\left[S_i^{(n)}\right]$$

$$= \frac{E[n_i]}{\mu_{2_i}} \rho_i E\left[W_{pull}^{q(i)}\right] \tag{3.13}$$

(iii) *Estimating* $E[S_i']$: Proceeding in a similar way and assuming the uniform property of Poisson,

$$E[S_i'] = \frac{E[n_i']}{\mu_{2_i}} \rho_j E\left[W_{pull}^{q(x)}\right] \tag{3.14}$$

The solution of equation (3.11) can be achieved by combining the results of equations (3.12)–(3.14) and using Cobham's iterative induction [17]. Finally, the new overall expected waiting time of the pull system ($E[\widehat{W_{pull}^q}]$) is achieved in the following manner:

$$E\left[W_{pull}^{q(x)}\right] = \frac{\sum_{i=1}^{max} \rho_i/\mu_{2_i}}{(1 - \sigma_{x-1})(1 - \sigma_x)} \tag{3.15}$$

$$E\left[\widehat{W_{pull}^q}\right] = \sum_{x=1}^{max} \frac{\lambda_x E\left[W_{pull}^{q(x)}\right]}{\lambda} \tag{3.16}$$

Thus, the expected access time, $E[T_{hyb-acc}]$, of our hybrid system is given by:

$$E[T_{hyb-acc}] = E[\mathscr{L}_{pull}] \sum_{i=1}^{K} \frac{S_i}{2} P_i + E\left[\widehat{W_{pull}^q}\right] \times \sum_{i=K+1}^{D} P_i \tag{3.17}$$

where according to packet fair scheduling, $s_i = [\Sigma_{i=1}^{M} \sqrt{\hat{P}_i \, l_i}]\sqrt{l_i/\hat{P}_i}$ and $\hat{P}_i = \frac{P_i}{\Sigma_{j=1}^{K} P_j}$.
The above expression provides an estimate of the average behavior of our hybrid
scheduling system.

3.3.2.2 Estimation of the Cutoff Value

One important system parameter that needs investigating is the cutoff point; that is the
value of K that minimizes the expected waiting time in the hybrid system. It is quite
clear from equations (3.9)–(3.17) that the dynamics of minimum expected waiting
time (delay) is governed by K. Furthermore, equation (3.17) has two components
for the minimum expected waiting time. While $\sum_{i=1}^{K} \frac{s_i}{2} P_i$ provides an estimate for
the delay accumulated from the push system, $E[W_{pull}^{q}] \times \sum_{i=K+1}^{D} P_i$ represents the
delay component arising from the pull system. For different values of K, these two
components change dynamically. Intuitively, for low values of K, most of the items
are pulled and the significant delay is accrued from the pulled items. The scenario gets
reversed for high values of K. It seems hard to derive a closed-form solution for the
optimal value of K. The cutoff point can be obtained algorithmically by estimating
both the delay components and overall expected delay at each iteration and preserving
the value of K that provides minimum expected delay. Alternatively to derive the
cutoff point, for a fixed value D, we analyze the pattern of the expected waiting time
with different values of K and fit the values to obtain a closed form equation of the
pattern. We have used *polynomial fit* with degree 3 to identify these patterns for three
different values of the access skew coefficient, $\theta = \{0.40, 0.80, 1.20\}$. This leads
to the equations for $E[T_{hybacc}] = f(K)$. For the sake of notational simplicity, we
use y to represent $E[T_{hybacc}]$. We first differentiate y with respect to K to get the
first derivative $\frac{\partial y}{\partial K}$. At the extreme points (maxima or minima) the derivative will
be 0. Hence, the expression for $\frac{\partial y}{\partial K}$ is made equal to 0 to get the extreme values
of K. As the polynomial is of degree 3, the slope of the curve $\frac{\partial y}{\partial K}$ is of degree 2.
Hence, two values of K are possible. We have taken only that value of K, that falls
in the range $0 \leq K \leq D$, as the minimum and maximum possible values of the
cutoff point are 0 and D, respectively. At this particular value of K, we compute
the value of y using the original equation. This is the minimum expected access
time with corresponding cutoff point for a particular value of θ. In order to check
the minima, we also computed the second order derivative with respect to K and
showed this derivative is positive (minimality condition) for that K. This is repeated
for $\theta = \{0.40, 0.80, 1.20\}$.

For example, the following three optimal values of K achieve the minimum waiting time for different values of θ and $D = 100$. When $\theta = 0.40$,

$$y = 27 \times 10^{-5} K^3 - 0.028 K^2 - 0.5K + 160$$

$$\left[\frac{\partial y}{\partial K}\right]_{\min\_y} = 81 \times 10^{-5} K^2 - 0.056K - 0.5 = 0$$

$$K = 77$$

$$y = 78.75191 \tag{3.18}$$

When $\theta = 0.80$

$$y = 13 \times 10^{-5} K^3 - 0.11 K^2 - 0.34 K + 100$$

$$\left[\frac{\partial y}{\partial K}\right]_{min\,\_y} = 39 \times 10^{-5} K^2 - 0.22 K - 0.34 = 0$$

$$K = 69$$

$$y = 66.875 \tag{3.19}$$

When $\theta = 1.20$

$$y = 0.01 K^2 - 4 \times 10^{-5} K^3 - 0.62 K + 55$$

$$\left[\frac{\partial y}{\partial K}\right]_{min\,\_y} = 0.02 K - 12 \times 10^{-5} K^2 - 0.62 = 0$$

$$K = 41$$

$$y = 43.633 \tag{3.20}$$

Figure 3.20 shows the variation of expected access time with different values of the size of the push set. The overall expected waiting time always achieves more or less a parabolic (bell-shaped) form with the global minima occurring at $K = \{77, 69, 41\}$ for $\theta = \{0.40, 0.80, 1.20\}$, respectively. The corresponding minimum expected waiting time is $\{79, 67, 44\}$ time units.

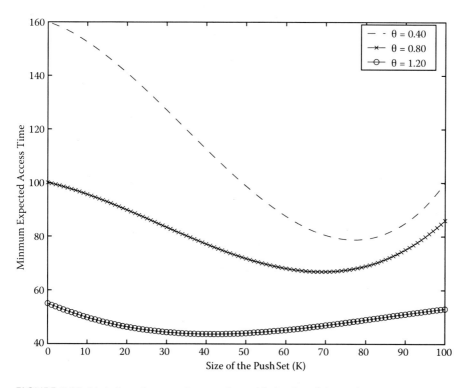

FIGURE 3.20 Variation of expected access time with the size of the push set.

3.4   EXPERIMENTAL RESULTS

In this section we validate our hybrid system by performing simulation experiments. The primary goal is to reduce the expected access time. We enumerate below the salient assumptions and parameters used in our simulation.

1. Simulation experiments are evaluated for a total number of $D = 100$ data items.
2. Arrival rate, $\lambda_{arrival}$, is varied between 5–20. The values of μ_1 and μ_2 are estimated as $\mu_1 = \sum_{i=1}^{K}(P_i \times L_i)$ and $\mu_2 = \sum_{i=K+1}^{D}(P_i \times L_i)$.
3. Length of data items is varied from 1 to 5. An average length of 2 is assumed.
4. Every client is assumed to have some priority randomly assigned between 1 and 5. These priorities are so defined that the lower the value, the higher the priority.
5. To keep the access probabilities of the items from being similar to very skewed, θ is dynamically varied from 0.20 to 1.40.
6. To compare the performance of our hybrid system, we have chosen four different hybrid scheduling strategies [18, 31, 32, 45] as performance benchmarks.

Figures 3.21 and 3.22, respectively, demonstrate the variation of the expected access time with different values of K and θ, for $\lambda = 10$ and $\lambda = 20$, in our hybrid

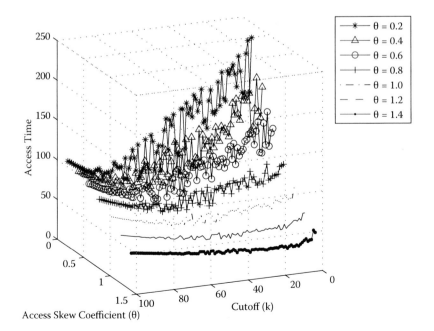

FIGURE 3.21   System behavior with $\lambda = 10$.

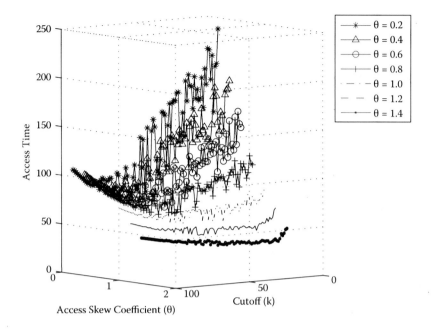

FIGURE 3.22 System behavior with $\lambda = 20$.

scheduling algorithm. In both cases, the expected access time is minimum (~ 40 time units) for high values of θ (~ 1.3) and moderate K. With decreasing values of K, the expected access time increases. This is because as K decreases, the number of items in the pull queue increases and those items take much more time to get serviced. On the other hand, the average access time also increases for very high values of K. This is because for pretty high K, the push set becomes very large and the system repeatedly broadcasts data items which are even not popular. Thus, the optimal performance is achieved when K is in the range 40–60.

Figure 3.23 shows the results of performance comparison, in terms of expected access time (in seconds), between our newly proposed hybrid algorithm with three existing hybrid schemes due to Su, et al. [45], Oh, et al. [32], and Guo, et. al. [18]. The plots reveal that our new algorithm achieves an improvement of $\sim 2 - 6$ secs. The possible reasons lie in the fact that while these existing scheduling algorithms use MRF and flat scheduling to select an item for transmission from the pull and push systems, our new algorithm uses the *stretch*, i.e., *max-request min-service-time* based scheduling and *packet fair scheduling* for pull and push systems, respectively. The effective combination of these two scheduling principles results in the lower expected access time in our hybrid scheduling algorithm.

To demonstrate the performance efficiency of the proposed hybrid scheduling, we have also looked into the minimum expected access time (for a particular K and θ) with different arrival rates (λ). The hybrid scheduling algorithm due to reference [31]

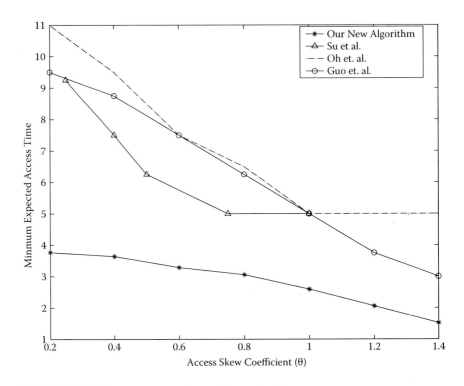

FIGURE 3.23 Performance comparison with varying skewness.

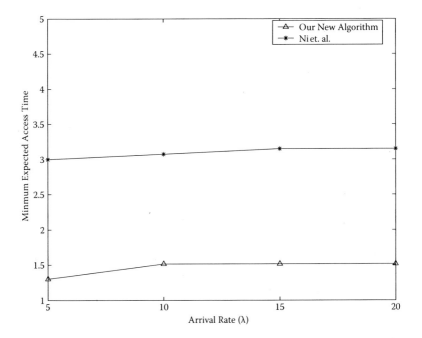

FIGURE 3.24 Performance comparison with different arrival.

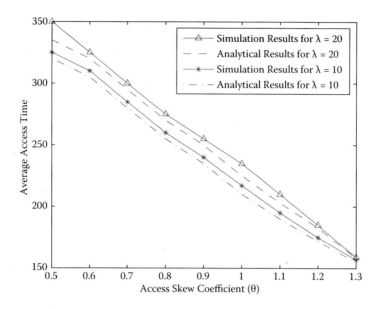

FIGURE 3.25 Simulation versus analytical results.

is chosen for comparison. Figure 3.24 shows that our algorithm consistently gains over existing hybrid scheduling [31] with different arrival rates. Note that the variation of expected access time with different arrival rates is pretty low. This also demonstrates the stability of our hybrid scheduling system.

Figure 3.25 depicts the comparative view of the analytical results, provided in equation (3.17), with the simulation results of our hybrid scheduling scheme. The analytical results closely match the simulation results for expected access time with almost $\sim 90\%$ and $\sim 93\%$ accuracy for $\lambda = 5$ and $\lambda = 20$, respectively. Thus, we can conclude that the performance analysis is capable of capturing the average system behavior with good accuracy. The little (~ 7–10%) difference exists because of the assumption of memoryless property associated with arrival rates and service times in the system.

Let us now investigate the dynamics of the cutoff point (K) achieved by our hybrid scheduling strategy. Figure 3.26 demonstrates that K lies in the range of 40–60 for three different arrival rates such as $\lambda = [5, 10, 20]$. Intuitively, this points out that the system has achieved a fair balance between push and pull systems, thereby efficiently combining both the scheduling strategies to achieve the minimum expected access time.

Figure 3.27 provides the comparison of the variation of optimal cutoff point provided by simulation and analytical results, for different values of access skew coefficient, θ. The plots point out that the simulation and analytical results of the optimal cutoff point closely match with a difference of only $\sim$ one percent to two percent.

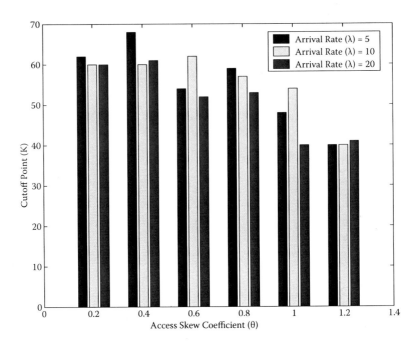

FIGURE 3.26 Variation of cutoff point (K).

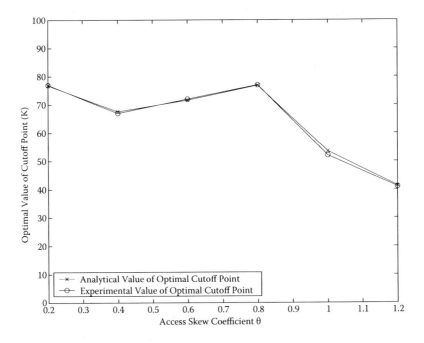

FIGURE 3.27 Simulation versus analytical results of the optimal cutoff point (K).

3.5 SUMMARY

In this chapter we proposed a new framework for hybrid scheduling in asymmetric wireless environments. The framework is initially designed for homogeneous, unit-length items. The push system operates on PFS and the pull part is based on MRF scheduling. The cutoff point used to separate push and pull systems is determined such that overall expected access delays are minimized. Subsequently, the system is enhanced to include the items of heterogeneous lengths. In order to take the heterogeneity into account, the pull part is now based on stretch-optimal scheduling. Performance modeling, analysis, and simulation results are performed to get an overall picture of the hybrid scheduling framework.

4 Adaptive Push-Pull Algorithm with Performance Guarantee

A dynamic hybrid scheduling scheme [37] is proposed, that does not combine the push and pull in a static, sequential order. Instead, it combines the push and pull strategies probabilistically depending on the number of items present and their popularity. In practical systems, the number of items in a push and pull set can vary. For a system with more items in the push set (pull set) than the pull set (push set), it is more effective to perform multiple push (pull) operations before one pull (push) operation.

The cutoff point, that is the separation between the push and the pull items, is determined in such a way that the clients are served before the deadline specified at the time of the request. In other words, the major novelty of our system is its capability to offer a performance guarantee to the clients. Once the analytic model has been devised, we take advantage of it to make our system more flexible. Because of the analytic model, system performance is already precisely known; it is possible to decide in advance if the value of K currently in use at the server will satisfy the client request on time. If not, K is updated in an appropriate way again looking at the system performance analysis.

4.1 ADAPTIVE DYNAMIC HYBRID SCHEDULING ALGORITHM

The strict sequential combination of push and pull fails to explore the system's current condition. In practical systems, it is a better idea to perform more than one push operation followed by multiple pull operations, depending on the number of items currently present in the system. As shown in Figure 4.1, the algorithm needs to be smart and efficient enough to get an accurate estimate of the number of continuous push and pull operations. Our proposed hybrid scheduling scheme performs this strategy based on the number of items present and their popularity.

We have assumed a single server, multiple clients, and a database consisting of D distinct items, of which K items are pushed and the remaining $(D - K)$ items are pulled. The access probability P_i of an item i is governed by the Zipf's distribution and depends on the access skew coefficient θ. When θ is small (value close to 0), P_i is well balanced but as θ increases (value close to 1), the popularity of the data items becomes skewed. From time to time the value of θ is changed dynamically for our hybrid system, that, in turn, results in dynamic variation of P_i and the size of the push and pull sets. PFS and MRF techniques are used for selecting the item to be pushed and pulled, respectively. After every push or pull operation, the next push or

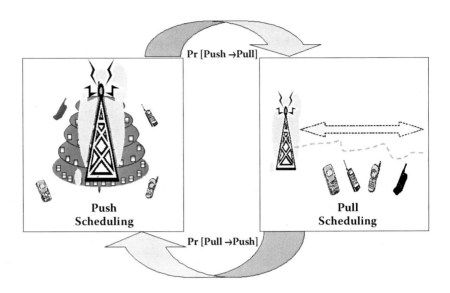

FIGURE 4.1 Probabilistic push-pull combination.

pull operation is probabilistically determined using the following equation:

$$\gamma_1 = Pr[push|push] = \frac{K}{D} \sum_{i=1}^{K} P_i$$

$$\gamma_2 = Pr[pull|push] = 1 - \gamma_1$$

$$\gamma_3 = Pr[pull|pull] = \frac{D - K}{D} \sum_{i=K+1}^{D} P_i$$

$$\gamma_4 = Pr[push|pull] = 1 - \gamma_3 \tag{4.1}$$

In other words, at the end of every push operation, the system checks if γ_1. If $\gamma_1 \geq Pr_1$ (where Pr_1 is a predefined value), the system goes for another push, or it switches to the pull mode. Similarly, at the end of every pull operation, it computes the value of γ_3. If $\gamma_3 \geq Pr_2$ (Pr_2 is predefined) then the system performs another pull operation, or it switches to the push mode.

At the server end, the system begins as a pure pull-based scheduler. If the request is for a push item, the server simply ignores the request as the item will be pushed according to the PFS algorithm sooner. However if the request is for a pull item, the server inserts it into the pull queue with the associated arrival time and updates its stretch value. Figure 4.2 provides the pseudo-code of the hybrid scheduling algorithm executing at the serverside.

4.1.1 ANALYTICAL UNDERPINNINGS

In this section we investigate the performance evaluation of our hybrid scheduling system by developing suitable analytical models. The arrival rate in the entire system

```
Procedure HYBRID SCHEDULING (Pr₁, Pr₂);
while true do
    begin
        1. select an item using PFS and push it;
        2. consider new arrival requests;
        3. ignore the requests for push item;
        4. append the requests for items in the
        pull queue;
        5. compute probabilities of γ₁ and γ₂
        6. if (Pr₁ <= γ₁) go to step 1
        7. or go to step 8
        8. if pull queue is not empty then
                9. use MRF to extract an item from pull
                queue;
                10. clear pending requests for that item;
                11. pull that particular item;
                12. consider new arrival requests;
                13. ignore the requests for push item;
                14. append the requests for items in
                pull queue;
        end-if
        15. compute probabilities of γ₃ and γ₄
        16. if (Pr₂ <= γ₃) go to step 8
        or go to step 1;
end-while
```

FIGURE 4.2 Hybrid scheduling algorithm at the server.

is assumed to obey the Poisson distribution with mean λ_1. The service times of both the push and pull systems are exponentially distributed with mean μ_1 and μ_2, respectively. The server pushes K items and clients pull the rest $(D - K)$ items. Thus, the total probability of items in push set and pull set are respectively given by $\sum_{i=1}^{K} P_i$ and $\sum_{i=K+1}^{D} P_i = (1 - \sum_{i=1}^{K} P_i)$, where P_i denotes the access probability of item i. We have assumed that the access probabilities P_i follow the *Zipf's distribution* with *access skew coefficient* θ, such that, $P_i = \frac{(1/i)^{\theta}}{\sum_{j=1}^{n}(1/j)^{\theta}}$. After every push the server performs another push with probability γ_1 and a pull with probability γ_2. Similarly, after every pull, it performs another pull with probability γ_3 and a push with probability γ_4.

Figure 4.3 illustrates the underlying birth and death process of our system, in which the arrival rate in the pull system is given by $\lambda = (1 - \sum_{i=1}^{K} P_i)\lambda_1$. Any state of the overall system is represented by the tuple (i, j), in which i represents the number of items in the pull system. On the other hand, j is a binary variable, with $j = 0$ (or 1) respectively representing whether the push system (or pull system) is currently being served by the server.

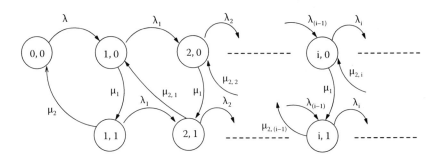

FIGURE 4.3 Performance modeling of hybrid system.

The arrival of a data item in the pull system results in the transition from state (i, j) to state $(i + 1, j)$, $\forall i$, such that $0 \leq i < \infty$ and $\forall j \in [0, 1]$. The service results in different actions. The service of an item in the push queue results in transition of the system from state $(i, j = 0)$ to state $(i, j = 1)$, with probability γ_2, $\forall i$ such that $0 \leq i < \infty$. With probability γ_1, the push service makes the system remain in the same state. On the other hand, the service of an item in the pull results in transition of the system from state $(i, j = 1)$ to the state $(i - 1, j = 0)$, with probability γ_4 and state $(i - 1, j = 1)$ with probability γ_3, $\forall i$, such that $1 \leq i < \infty$. The state of the system at $(i = 0, j = 0)$ represents that the pull queue is empty and any subsequent service of the elements of push system leaves the system in the same $(0, 0)$ state. Obviously, the state $(i = 0, j = 1)$ is not valid because the service of an empty pull queue is not possible.

In the steady state, using the flow balance conditions of *Chapman-Kolmogrov's* equation [17], we have the following three equations representing the system's behavior:

$$p(i, 0) = \frac{p(i - 1, 0)\lambda + p(i + 1, 1)\gamma_4\mu_2}{(\lambda + \gamma_2\mu_1)} \tag{4.2}$$

$$p(i, 1) = \frac{p(i, 0)\gamma_2\mu_1 + p(i - 1, 1)\lambda}{(\lambda + \gamma_3\mu_2 + \gamma_4\mu_2)} \tag{4.3}$$

$$p(0, 0)\lambda = p(1, 1)\mu_2 \tag{4.4}$$

where $p(i, j)$ represents the probability of state (i, j). Although the first two equations represent the overall behavior of the system for push (upper chain) and the pull system (lower chain), the last equation actually represents the initial condition of the system. The most efficient way to solve the above equations is using *z transforms* [17]. Performing z transforms of equation 4.2 and equation 4.3 and using the initial condition, we get a pair of transformed equations:

$$P_2(z)\gamma_4\mu_2 = z[P_1(z) - p(0, 0)](\lambda + \gamma_2\mu_1) - z^2\lambda P_1(z) + p(1, 1)\gamma_4\mu_2 \tag{4.5}$$

$$P_2(z) = \frac{\gamma_2\mu_1[P_1(z) - p(0, 0)]}{(\lambda + \gamma_3\mu_2 + \gamma_4\mu_2 - z\lambda)} \tag{4.6}$$

Now, estimating the system behavior at the initial condition, we can state the following normalization criteria: The occupancy of pull states is the total traffic of pull queue and is given by: $P_2(1) = \sum_{i=1}^{C} p(i, 1) = \rho$. The occupancy of the push states (upper chain) is similarly given by: $P_1(1) = \sum_{i=1}^{C} p(i, 0) = (1 - \rho)$. Using these two relations in equation (4.5), we can obtain the initial probability, $p(0, 0)$. The initial probability of the system, i.e., probability of an empty pull queue is given by the following equation:

$$p(0, 0) = \frac{\mu_1(\gamma_2 - \gamma_2\rho - \rho\gamma_4\mu_2)}{\lambda + \gamma_2\mu_1 - \gamma_4\lambda} \tag{4.7}$$

Generalized solutions of equations (4.5) to obtain all values of probabilities $p(i, j)$ become very complicated. Thus, the best possible way is to go for an expected measure of system performance, such as the average number of elements in the system and average waiting time. The most convenient way to obtain this expected system performance is to differentiate the z transformed equation (equation (4.5)), and capture the values of the z transformed variable at $z = 1$.

$$\gamma_4\mu_2 \frac{dP_2(z)}{dz}\Big|_{z=1} = \gamma_2\mu_1 \frac{dP_1(z)}{dz}\Big|_{z=1} + (1 - \rho)(\gamma_2\mu_1 - \lambda) - p(0, 0)(\lambda + \gamma_2\mu_1)$$

$$E[\mathcal{L}_{pull}^q] = \frac{dP_2(z)}{dz}\Big|_{z=1} \tag{4.8}$$

where $\frac{dP_1(z)}{dz}\Big|_{z=1}$ gives the number of elements in push system in PFS. Once, we have the expected number of items in the pull system from equation (4.8), using Little's formula [17], we can easily obtain the estimates of average waiting time of the system ($E[W_{pull}]$), and expected number of items ($E[\mathcal{L}_{pull}^q]$) in the pull queue as:

$$E[W_{pull}^q] = E[W_{pull}] - \frac{1}{\mu_2} = \frac{E[\mathcal{L}_{pull}]}{\lambda} - \frac{1}{\mu_2} \tag{4.9}$$

If K represents the number of items in the push system, then the expected cycle time of the push system is given by: $\sum_{i=1}^{K} \frac{S_i P_i}{(1-\rho)\mu_1}$. Hence, the expected access time ($E[T_{hyb-acc}]$) of our hybrid system is given by:

$$E[T_{hyb-acc}] = \sum_{i=1}^{K} s_i P_i + E[W_{pull}^q] \times \sum_{i=k+1}^{D} P_i \tag{4.10}$$

where according to the packet fair scheduling, $s_i = \frac{\sum_{j=1}^{K} \sqrt{\hat{P}_j}}{\sqrt{\hat{P}_i}}$ and $\hat{P}_i = \frac{P_i}{\sum_{j=1}^{K} P_j}$. The above expression provides an estimate of the average behavior of our hybrid system.

4.1.2 SIMULATION EXPERIMENTS

In this section we perform the experiments to demonstrate the performance efficiency of our hybrid system. In order to compare the performance of our hybrid system, we have chosen our previous hybrid scheduling algorithm [34] as performance benchmarks. The prime goal of the entire scheme is to reduce the expected access time.

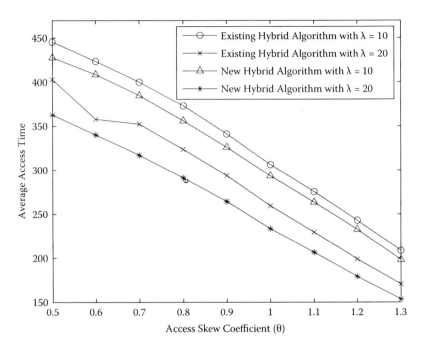

FIGURE 4.4 Improvement in average access time.

Before going into the details of the simulation results, we enumerate the assumptions
we have used in our simulation.

1. The simulation experiments are evaluated for $D = 1000$ items. The system
   performs a push and pull operation in a reciprocal manner, unless the pull
   queue is empty.
2. In order to remain consistent with the analysis, the arrival and service rates
   are assumed to obey Poisson distribution. The average value of arrival rate
   (λ) is taken to be 10 and 20. The average value of service rates (push and
   pull), μ_1 and μ_2 are assumed to be 1.
3. In order to keep the access probabilities of the items from similar to very
   skewed, θ is dynamically varied from 0.50 to 1.50.

Figure 4.4 demonstrates the variation of expected access time with different values of θ, for arrival rates of 10 and 20, respectively, in our hybrid scheduling algorithm, for 1000 items. Note that, in both cases, the expected access time for our new
hybrid scheduling is sufficiently lower than the expected access time for existing hybrid scheduling. The prime reason behind this is that the hybrid scheduling captures
the requirement of the system by probabilistically combining push- and pull-based
scheduling principles. Figure 4.5 shows that the hybrid scheduling achieves a cutoff
point in the range 360–430 and 360–460, respectively, for arrival rates of 10 and 20
with 1000 data items. This explains the reason that our hybrid scheduling makes a fair

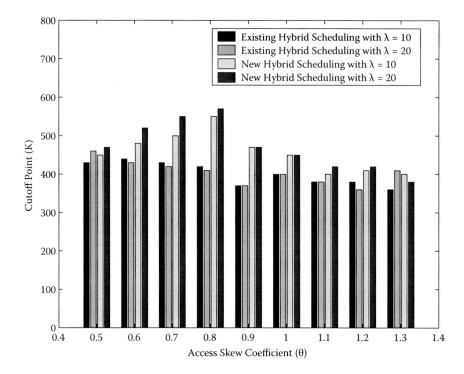

FIGURE 4.5 Dynamics of cutoff point.

combination of both push and pull systems, that is required to improve the expected access time. Figure 4.6 depicts the comparative view of the analytical results with the simulation results, for 1000 data items. The analytical results closely match with the simulation results for expected access time with almost $\sim 90\%$ and $\sim 93\%$ accuracy for $\lambda = 10$ and $\lambda = 20$, respectively. Thus, we can conclude that the performance analysis is capable of capturing the average system behavior with good accuracy. The little ($\sim 7\text{--}10\%$) differences exist because of the assumption of memory less property associated with arrival rates and service times in the system.

4.2 PERFORMANCE GUARANTEE IN HYBRID SCHEDULING

While the performance evaluation of our new basic hybrid scheduling algorithm has already pointed out its significant gains in both response time and minimizing the value of the cutoff point when compared to existing algorithms and pure push-based scheduling, we proceed to the complete version of our new algorithm [36] with performance guaranteed quality that our new scheduling scheme is capable of offering. Such a performance guarantee is required to deliver, for example, the wireless voice and data packets within a precise timeframe of service, thereby ensuring a certain level of quality-of-service (QoS).

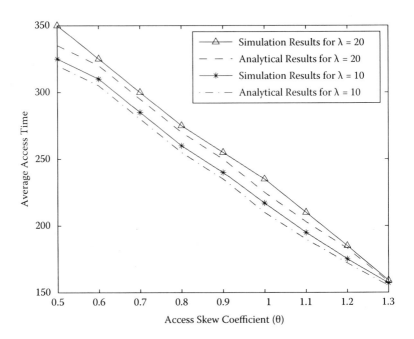

FIGURE 4.6 Comparison of analytical and simulation results.

Essentially for all D possible values of the cutoff point, the server computes the expected hybrid waiting time $E[T_{hyb-access}]$. From now on, let $E[T_{hyb-access}(i)]$ denote such expected hybrid waiting time when i is the cutoff point value. These D expected waiting times are stored, one at a time along with the index of the cutoff point that generates it, sequentially in a vector V. That is, for all $1 \leq i \leq D$, $\mathbf{V}[i, 1] = [E[T_{hyb-access}(i)]$ and $V[i, 2] = i]$. Moreover V is maintained sorted with respect to the first component, i.e., the expected hybrid waiting time. With this structure, the server can extract the first element $\mathbf{V}[0]$ of this vector in a single access, which will indicate in correspondence which value K_0 of the cutoff point, the minimum expected hybrid waiting time $V[0, 1] = E[T_{hyb-access}(K_0)]$ is achieved. The server broadcasts $V[0]$ from time to time, thereby informing the clients of the best performance it can provide. Moreover, the server continuously broadcasts the basic hybrid scheduling that corresponds to the cutoff point K in use we discussed in the previous section. On the other side, when a client sends a request for any item j, it also specifies an expectation $\Delta(j)$ of its possible waiting time for item j. Indeed, $\Delta(j)$ reflects the nature of the application, and the tolerance of the client. For example, a client requesting for any real-time video application, will expect a time much lower than any client requesting data service specific applications. Moreover, an impatient client could ask that its request is served in a time much lower than its moderate counterparts.

To accept a client request for item j with expectation $\Delta(j)$, the server estimates the expected waiting time for j at that instant using the values stored in vector V and the knowledge of the current cutoff point K in use for the hybrid scheduling algorithm broadcast by the server. If the expected hybrid waiting time provided by the system is smaller or equal to the expectation time of the client's request, then the certain level of QoS expected by the client is guaranteed. Otherwise, the server checks whether the item j belongs to its current push set, that is if $j \leq K$. If this is true, it compares the client request deadline with the expected waiting time guaranteed by the packed fair scheduling queueing for the push part of the system, say $E[T_{PFS}(j)]$. Recall that such a value is known and is proportional to the space S_j between two instances of j in the packed fair queueing scheduling [19], and it is different from the overall expected waiting time of the server although it depends on the cutoff point in use. If the expectation of the client $2E[T_{PFS}(j)]$ is smaller than or equal to $\Delta(j)$, the request can still be accepted and the performance guaranteed. Note that the $E[T_{PFS}(j)]$ is doubled to make allowance for the fact that the system pulls one item between two consecutive pushed items. Otherwise, the request can be accepted only if the cutoff point is updated. Indeed, the server will perform a binary search on the vector V to look for a cutoff point value whose corresponding expected hybrid waiting time is the largest value smaller than or equal to $\Delta(j)$. Note that such a value always exists if $V[0, 1] \leq \Delta(j)$. Then, the cutoff point is updated accordingly and the scheduling reinitialized. Note that the adjustment in push and pull sets results in some overheads, and in practice, the server may be forced to reject requests to avoid to pay such an overhead too frequently. Figure 4.7 provides a pseudo-code for this entire procedure of performance guarantee.

Performance Guarantee in Hybrid Scheduling

```
1. for (i = 1 to D) do
2. compute the average waiting time E[T_hyb-access(i)];
3. sort all the values of E[T_hyb-access(1)], ..., E[T_hyb-access(D)] and
   store them in increasing order in a vector V
4. broadcast to the clients the min{affordable waiting time} V[0];
do
5. accept the client's request for any item j with expected
      waiting-time Δ(j) ≥ E[T_hyp-access(K)], where K = current cutoff;
6. if (condition at line 5 is not verified and j ≤ K)
         accept the client's request for any item j with expected
         waiting time Δ(j) ≥ 2E[T_PFS(j)], where E[T_PFS(j)] is the
         expected waiting time guaranteed by packet fair scheduling;
7. if (both conditions at lines 5 and 6 are not verified) and
         (Δ(j) is larger than expected waiting time stored in V[0, 1])
8. get the largest value of E[T_hyb-access(j)] ≤ Δ(j) from V
9. adjust the cutoff point and restart new hybrid scheduling;
10. otherwise reject the request.
while (true)
```

FIGURE 4.7 Algorithm for performance guarantee in hybrid scheduling.

4.3   SUMMARY

In this chapter we have improved our hybrid scheduling framework to make it adaptive to the system's behavior and provide a certain level of performance guarantee. Instead of strict, sequential push and pull operations, the hybrid scheduling framework now probabilistically determines the number of consecutive push and pull operations based on the system's requirements. Subsequently, we propose a strategy to provide a certain level of performance guarantee by meeting the clients' deadlines.

5 Hybrid Scheduling with Client's Impatience

In most practical systems, clients often become impatient while waiting for the designated data item. After a tolerance limit, the client may depart from the system, resulting in a drop of access requests. This behavior significantly affects the system performance, which needs to be properly addressed. There are also ambiguous cases that reflect the false situation of the system. Consider the scenario in which a client gets impatient and sends multiple requests for a single data item to the server, as shown in Figure 5.1. Even if that particular data item is not requested by any other client, its access probability becomes higher. In existing systems, the server remains ignorant of this fact and thus considers the item as popular and inserts it into the push set or pull it at the expense of some other popular item. In contrast, our work [43] reduces the overall waiting time of the system taking care of such anomalies.

5.1 HYBRID SCHEDULING ALGORITHM

The major novelty of our strategy lies in its consideration for clients' impatience that is incorporated in two different ways, leading to two different strategies as shown in Figure 5.2. Although the basics of both strategies are similar, the first one considers that the impatience of a client results in a departure from the system. This strategy is termed as *hybrid scheduling with clients departure*. Whereas, the second strategy considers the fact that a client's impatience compels it to send spurious requests for a particular data item, thereby creating an anomalous (ambiguous) situation in the system. We term this strategy as *hybrid scheduling with anomalies*.

In general, the system begins with operating as a pure pull system providing on-demand service for every client. When the number of client's access request rate increases and broadcasting the same item to different clients causes downstream bandwidth wastage, the algorithm shifts to the hybrid mode. The items are now divided into two disjoint sets: the push set of cardinality K and the pull set of cardinality $D - K$. The items to be pushed are governed by flat round-robin scheduling as shown in Figure 5.3.

On the other hand, the item that maximizes stretch (max-request min-service time) is selected to be pulled by the server. Every push is followed by a pull, provided that the pull set (queue) is not empty. If there are no items in the pull queue, then the server simply continues pushing the items using a flat schedule. However, after transmitting each page, the server attempts λ more access requests arriving into the system. If the request is for a push item, the server simply ignores the request as the item would be pushed anyway according to the broadcast schedule.

If the request is for a pull item, then the server first checks whether the request is for a new item or an already requested item. If it is for a new item, the item is inserted

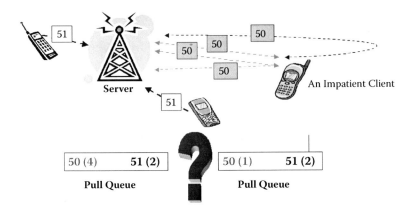

FIGURE 5.1 Clients' impatience: spurious requests from an impatient client.

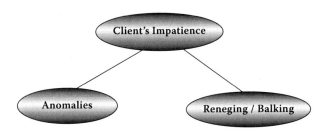

FIGURE 5.2 Effects of clients' impatience.

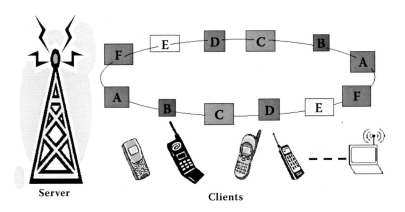

FIGURE 5.3 Flat round-robin scheduling for push items.

```
Hybrid scheduling with clients' departure;
while true do
begin
Broadcast all the pages of an item, selected
according to the flat scheduling;
After broadcasting each page
Take-Access-with Drop();
if the pull queue is not empty then
   extract an item from the pull queue
   that optimizes the stretch;
   clear the number of pending requests for
   that item and pull it;
Take-Access-with-Drop() /*procedure call */
end;
```

FIGURE 5.4 Hybrid scheduling with clients' departure.

into the pull queue and its stretch value is calculated. Next, the server checks for the client's impatience and tolerance. The impatience is considered in the two strategies as follows.

5.1.1 HYBRID SCHEDULING WITH CLIENTS' DEPARTURE

If the request is for an existing item, the server checks whether one or more clients are getting impatient and losing their tolerance limit. Anticipating departures of such clients, the server drops their requests and stores their previous waiting time (departure time − arrival time). It then updates the stretch value of the data items in the pull queue considering only the request of existing clients that are not impatient. A pseudo-code of the strategy is depicted in Figure 5.4. The procedure *take access-with drop()* considers λ more requests, processes them, and inserts in the pull queue, after considering the number of requests dropped due to the client's departure. A pseudo-code of this procedure is shown in Figure 5.5.

5.1.2 HYBRID SCHEDULING WITH ANOMALIES

While considering a request for an item that is already in the pull queue, the server checks for anomalies arising from spurious requests of impatient client(s) for a particular data item. While exceeding the tolerance limit, a single client can send a large number of requests for a particular data item, thereby making it *pseudo-popular*. In existing hybrid scheduling schemes, the server is ignorant of this fact and considers the item as a popular one, even if it is requested by a single client. In order to remove this anomaly, the server now considers only unique requests for data items, i.e., if

> **Procedure: Take-Access-with-Drop();**
> ```
> Take λ more accesses;
> ```
> **if** the request is for push items **then**
>   ignore the requests;
> **if** the request is for pull items **then**
>   Compute the number of impatient clients
>   leaving and number of clients remaining;
>   insert the request for the pull item into
>   the pull queue (with arrival time);
>   update the stretch value of data items in
>   pull queue based on number of remaining
>   clients;

FIGURE 5.5 Take access with drop requests.

the request is from a new client, and not from the same client(s) who have already requested this item before, as depicted in Figure 5.6. Thus, the system computes the unique requests by the clients, i.e., the effective number of requests for data item i. The stretch values of the items in the pull queue are now updated using these unique requests.

A pseudo-code of this algorithm is shown in Figure 5.7. The procedure *proc-req-anomalies()* takes λ more requests, processes the requests, and inserts them into the pull queue after removing the anomalies associated with the requests. The pseudo-code of this procedure is shown in Figure 5.8

The dynamics of the system often lead to changes in the arrival rate of the access requests, in other words, in the access skew coefficient (θ). Hence, the access probabilities for all data items are recalculated. Based on these new access probabilities,

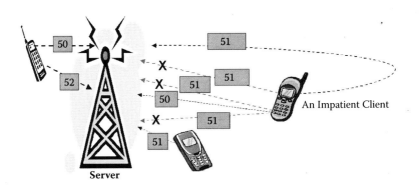

FIGURE 5.6 Considering unique requests from clients.

```
Hybrid scheduling with anomalies;
while true do
begin
Broadcast all the pages of an item,
according to the flat scheduling;
After broadcasting each page
ttProc-Req-Anomalies();
if the pull queue is not empty then
    extract an item from the pull queue
    that optimizes the stretch;
    clear the number of pending requests for
    that item and pull it;
Proc-Req-Anomalies(); /* procedure call */
end;
```

FIGURE 5.7 Hybrid scheduling with anomalies.

the cutoff point (K) is calculated dynamically. This needs dynamic shuffle of some items between the push- and the pull-set. Whenever a client requires an item, it sends a request for that item to the server. The clients can request any item from the server. No matter whether the item is currently being broadcast or disseminated, the client simply passes its request for the interested item to the server and listens to the channel until its desired item is obtained. This procedure is highlighted in Figure 5.9.

```
Procedure: Proc-Req-Anomalies();
Take λ accesses;
if the request is for push items then
  ignore the requests;
if the request is for pull items then
  if the same item is not requested
  by same client(s)
    insert the request for this pull item into
    the pull queue (with arrival time);
    update the stretch value of the data
    items in pull queue;
```

FIGURE 5.8 Process requests with anomalies.

```
Procedure CLIENT-REQUEST (i):
/* i : item the client is
interested in */
begin
send to the server the request for
item i;
wait until listen for i on the
channel
end
```

FIGURE 5.9 Algorithm at the client side.

5.2 PERFORMANCE MODELING AND ANALYSIS

In this section we analyze the performance of our hybrid scheduling algorithm. Recall that we proposed two different schemes to incorporate a client's impatience and accordingly we analyze the system performance by developing two different queuing models. However, the primary goal of both the analysis is to estimate the minimum expected waiting time (delay) of the hybrid system. Before proceeding further, let us enumerate the parameters and assumptions used.

5.2.1 ASSUMPTIONS

1. The arrival rate in the entire system is assumed to obey the Poisson distribution with mean λ'. This includes the arrival rate in both push and pull systems. Although the arrival rate of the push system is assumed fixed, the departure of impatient clients or their spurious requests changes the arrival rate of the pull system at every step. The initial arrival rate of the pull system is assumed to be λ. The pull queue contains data items that are yet to be served. Thus by the term pull system, we mean the items waiting in pull queue, together with the item(s) currently getting service.

2. The service times of both the push and pull systems are exponentially distributed. Again, the mean service time of the push system is fixed; however, the clients' impatience changes the service time of the pull system. We represent the initial service time of pull system by μ_2.

3. Let C, D, and K respectively represent the maximum number of clients, total number of distinct data items, and the cutoff point. The server pushes K items while clients pull the remaining $(D - K)$ items. Thus, the total probability of items in the push and pull sets are respectively given by $\sum_{i=1}^{K} P_i$ and $\sum_{i=K+1}^{D} P_i = (1 - \sum_{i=1}^{K} P_i)$, where P_i denotes the access probability of item i. Basically, it gives a probabilistic measure of item's popularity among the clients. We have assumed that the access probabilities follow the *Zipf's distribution* with *access skew coefficient* θ, such that

TABLE 5.1
Symbols Used for Performance Analysis

Symbols	Meanings
D	Maximum number of data items
C	Maximum number of clients
i	Candidate data item
K	Cutoff point separating push and pull sets
P_i	Access probability of item i
L_i	Length of item i
λ'	Overall system arrival rate
λ	Initial arrival rate in pull queue
μ_1	Push queue service rate
μ_2	Initial service rate in pull queue
$E[W_{pull}]$	Expected waiting time of pull system
$E[W^q_{pull}]$	Expected waiting time of pull queue
$E[\mathscr{L}_{pull}]$	Expected number of items in the pull system
$E[\mathscr{L}^q_{pull}]$	Expected number of items in the pull queue

$P_i = \frac{(1/i)^\theta}{\sum_{j=1}^{n}(1/j)^\theta}$. Items are numbered from 1 to D and are arranged in the decreasing order of their access probabilities, i.e., $P_1 \geq P_2 \geq \ldots \geq P_D$. Table 5.1 lists the symbols with their meaning used in the context of our analysis.

Let us now analyze the system performance for achieving the minimal waiting time. First, we discuss the model when the client loses its patience and leaves the system. Next, we discuss the system where an impatient client transmits spurious requests for a particular data item. As mentioned, this situation creates anomaly in the system, and the server needs to ignore such requests.

5.2.2 CLIENT'S DEPARTURE FROM THE SYSTEM

Here we assume that a client's impatience results in its departure from the system before the item is actually serviced. This impatience generally takes two forms [17]: (1) the reluctance of the customer to remain in the system is known as *reneging*; (2) excessive reluctance might restrain the customer to even join the system, which is known as *balking*. These two behaviors significantly affect the arrival/service rate and average system performance. In our analysis, we have assumed the duration of the waiting time of a client (before leaving) to follow exponential distribution with mean $1/\tau$. If $\bar{\lambda}_m$ represents the request arrival rate for mth data item, then $\bar{\lambda}_m = P_m \lambda$, where λ is the initial request arrival rate of the entire pull system. If the request arrives at time t and does not depart the system before servicing the mth data at time Γ, then

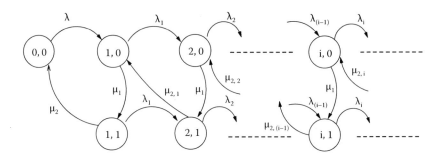

FIGURE 5.10 Performance modeling of our hybrid system.

expected number of requests, $E[R_i]$, satisfied by transmission of mth item is given by:

$$E[R_m] = \int_0^\Gamma \bar{\lambda}_m e^{-\tau(\Gamma-t)} dt$$
$$= \frac{P_m \lambda}{\tau}(1 - e^{-\tau\Gamma}) \tag{5.1}$$

Also, for Poisson arrival, the expected number of requests arriving in time period Γ is given by $\lambda\Gamma$. Thus, the expected number of drop requests, $E[R_d]$, is measured as:

$$E[R_d] = \lambda\Gamma - E[R_m]$$
$$= \lambda\Gamma - \frac{P_m \lambda}{\tau}(1 - e^{-\tau\Gamma}) \tag{5.2}$$

Our next objective is to estimate the expected waiting time of our hybrid system considering the clients balking and reneging due to client's impatience.

Figure 5.10 illustrates the birth and death model of our system. For any variable i, the ith state of the overall system is represented by the tuple (i, j), where i represents the number of items in the pull system and $j = 0$ (or 1) respectively represents whether the push system (or pull system) is being served. The arrival of a data item in the pull system results in the transition from state (i, j) to state $(i + 1, j)$, $\forall i \in [0, \infty]$, and $\forall j \in [0, 1]$. The service of an item results in transition of the system from state $(i, j = 0)$ to state $(i, j = 1)$, $\forall i \in [0, \infty]$. On the other hand, the service of an item in the pull results in transition of the system from state $(i, j = 1)$ to the state $(i - 1, j = 0)$, $\forall i \in [1, \infty]$. Note that, the arrival and service rates in the pull system are both different at each state. Naturally, the state of the system at $(i = 0, j = 0)$ represents that the pull queue is empty and any subsequent service of the elements of the push system leaves the system in the same $(0, 0)$ state. Obviously, state $(i = 0, j = 1)$ is not valid because the service of an empty pull queue is not possible. The arrival rates at different states are now represented by $\lambda_0, \lambda_1, \ldots, \lambda_i, \ldots,$ where $\lambda_0 = \lambda$. Furthermore, λ_i is different from $\bar{\lambda}_m$ discussed before. While $\bar{\lambda}_m$ represents the request arrival rate for mth data item, λ_i denotes the total arrival rate of requests for all i items present in the system, i.e., $\lambda_i = \sum_{m=0}^i \bar{\lambda}_m$. Similarly, the service rates at different states are denoted by $\mu_{2,j}$, where $1 \le j \le n$ and $\mu_{2,1} = \mu_2$.

In the steady-state, using the flow balance conditions of Chapman–Kolmogrov's equation [17], the initial system behavior is represented by:

$$p(0, 0)\lambda = p(1, 1)\mu_2 \tag{5.3}$$

where $p(i, j)$ represents the probability of state (i, j). The overall behavior of the system for push (upper chain in Figure 5.10) and the pull system (lower chain) are given by the following two generalized equations:

$$p(i, 0)(\lambda_i + \mu_1) = p(i - 1, 0)\lambda_{i-1} + p(i + 1, 1)\mu_{2,i+1} \tag{5.4}$$

$$p(i, 1)(\lambda_i + \mu_{2,i}) = p(i, 0)\mu_1 + p(i - 1, 1)\lambda_{i-1} \tag{5.5}$$

Balking [17] is generally estimated by using a series of monotonically decreasing functions of the system size multiplied by the initial arrival rate, λ. If b_i is the balking function at ith state, then $\lambda_i = b_i\lambda$, where $0 \leq b_{i+1} \leq b_i \leq 1, (\forall i > 0, b_0 = 1)$. The most practical discouragement (balking) function is $b_i = e^{-i\alpha}$, where α is a constant. This takes the queue size into account and discourages customers from joining in large-sized queues. However, in practical systems, the discouragement does not always arrive from excessive queue sizes. These customers might instead join the system and continuously retain the prerogative to *renege* if the waiting time is intolerable. This reneging function $r(i)$ [17] at ith state is defined by:

$$r(i) = \lim_{\Delta t \to 0} \frac{Pr[\text{unit reneges during } \Delta t]}{\Delta t} \tag{5.6}$$

The service rate of pull queue now takes the form: $\mu_2 = \mu_2 + r(i)$. A good possibility of the reneging function is: $r(i) = e^{i\alpha/\mu_2}$. Note that both balking and reneging functions are assumed to follow exponential distribution, which is in accordance with the distribution obeyed by request's waiting time. From equations (5.4) and (5.5), we get:

$$p(i, 0)(e^{-\alpha i}\lambda + \mu_1) = p(i - 1, 0)\lambda e^{-\alpha(i-1)} + p(i + 1, 1)\mu_2 + e^{(i+1)\frac{\alpha}{\mu_2}}$$

$$p(i, 1)\lambda e^{-\alpha i} + p(i, 1)\mu_2 + p(i, 1)e^{\alpha\frac{i}{\mu_2}} = p(i, 0)\mu_1 + p(i - 1, 1)e^{-\alpha(i-1)} \tag{5.7}$$

The most efficient way to solve equation (5.7) is using *z transforms* [17]. From the definition of z transforms, the resulting solutions are of the form:

$$P_1(z) = \sum_{i=0}^{C} p(i, 0)z^i \quad \text{and} \quad P_2(z) = \sum_{i=0}^{C} p(i, 1)z^i \tag{5.8}$$

Using subsequent z transforms, equation (5.7) yields:

$$\lambda\left[P_1\left(\frac{z}{e^\alpha}\right) - p(0, 0)\right] + \mu_1[P_1(z) - p(0, 0)]$$

$$= \lambda z\left[P_1\left(\frac{z}{e^\alpha}\right)\right] + \frac{1}{z}[P_2(z) - p(0, 1) - p(1, 1)]$$

$$+ \frac{1}{z}\left[P_2\left(ze^{\frac{\alpha}{\mu_2}}\right) - p(0, 1) - p(1, 1)\right] \tag{5.9}$$

Similarly, transforming equation (5.7) leads to:

$$\lambda P_2\left(\frac{z}{e^\alpha}\right) + P_2\left(ze^{\frac{\alpha}{\mu_2}}\right) = \mu_1 P_1(z) - p(0,0) + zP_2\left(\frac{z}{e^\alpha}\right) \qquad (5.10)$$

Now, putting $z = 1$ in equation (5.9), we can obtain the probability $p(0,0)$ of the idle state as:

$$\lambda\left[P_1\left(\frac{1}{e^\alpha}\right) - p(0,0)\right] + \mu_1[P_1(1) - p(0,0)]$$

$$= \lambda\left[P_1\left(\frac{1}{e^\alpha}\right)\right] + \mu_2[P_2(1) - p(1,1)] + P_2\left(e^{\frac{\alpha}{\mu_2}}\right) - p(1,1)$$

$$p(0,0) = \frac{\mu_2\rho - \mu_1(1-\rho) + \frac{\rho}{1-e^{\frac{\lambda}{\mu_2}}}}{\frac{\lambda}{\mu_2} - \mu_1} \qquad (5.11)$$

Deriving closed form solutions of equations (5.9) and (5.10) to evaluate the state probabilities seems impossible. Instead, we measure the expected performance of the overall system. In order to estimate the average number of items in the pull system, equation (5.9) is differentiated (at $z = 1$). Now, the occupancy of push and pull states are respectively given by $P_1(1) = \sum_{i=0}^\infty p(i,0) = 1 - \rho$ and $P_2(1) = \sum_{i=0}^\infty p(i,1) = \rho$, where $\rho = \frac{\lambda_{eff}}{\mu_{eff}} = \frac{\sum_{i=0}^\infty \lambda_i p(i,1)}{\sum_{i=1}^\infty \mu_{2,i} p(i,1)}$. Differentiating equation (5.9) and using these values of $P_1(1)$ and $P_2(1)$, we get:

$$\mu_2\frac{dP_2(z)}{dz} + \frac{dP_2}{dz}\left(ze^{\frac{\alpha}{\mu_2}}\right) = \mu_1 P_1(1) + \mu_1\frac{dP_1}{dz}$$

$$-(\lambda+\mu_1)\frac{\mu_1\rho - \mu_1(1-\rho) + \frac{1}{1-e^{\frac{\alpha}{\mu_2}}}}{\lambda/\mu_2 - \mu_1} - \lambda P_1(1/e^\alpha) - \lambda\frac{dP_1}{dz}\left(\frac{1}{e^\alpha}\right)$$

$$+ 2\lambda P_1(1/e^\alpha) + \lambda P_1(1/e^\alpha)E[\mathscr{L}_{pull}] = \frac{dP_2(z)}{dz}|_{z=1} = \left(\mu_1 + \frac{1}{1-e^{\frac{\alpha}{\mu_2}}}\right)^{-1}$$

$$\times\left[\mu_1\rho + \mu_1 E[\mathscr{L}_{push}] - (\mu_1+\lambda)\frac{\mu_1\rho - \mu_1(1-\rho) + \frac{\rho}{1-e^{\frac{\alpha}{\mu_2}}}}{\frac{\lambda}{\mu_2} - \mu_1}\right]$$

$$+ \lambda E[\mathscr{L}_{push}]e^{\alpha/mu_2}, \quad \left(\text{where } E[\mathscr{L}_{push}] = \frac{dP_1(z)}{dz}|_{z=1}\right) \qquad (5.12)$$

Once we have the expected number of items in the pull system from equation (5.12), using Little's formula [17], we can easily estimate the average waiting time of the system ($E[W_{pull}]$), average waiting time of the pull queue ($E[W_{pull}^q]$), and expected number of items ($E[\mathscr{L}_{pull}^q]$) in the pull queue as follows:

$$E[W_{pull}] = \frac{E[\mathscr{L}_{pull}]}{\lambda}, \quad E[\mathscr{L}_{pull}^q] = E[\mathscr{L}_{pull}] - \frac{\lambda}{\mu_2} \text{ and } E[W_{pull}^q] = E[W_{pull}] - \frac{1}{\mu_2}$$

Since the push system is governed by flat scheduling, the average cycle time of the push system is given by: $\frac{K}{2(1-\rho)\mu_1}\sum_{i=1}^K P_i$. Thus, the overall minimum expected access time, $(E[T_{hyb-acc}]$, of our hybrid system is:

$$E[T_{hyb-acc}] = \frac{K}{2(1-\rho)\mu_1} \sum_{i=1}^{K} P_i + E[W_{pull}] \sum_{i=K+1}^{D} P_i \qquad (5.13)$$

This gives a suitable measure of the performance of our hybrid, heterogeneous system when the clients get impatient and leave the system at certain intervals. Our next objective is to analyze the performance of the system, when the impatience does not force the clients to leave the system, but makes them to transmit spurious requests for the same data item.

5.2.3 Anomalies from Spurious Requests

As discussed earlier, the anomaly arises from the clients making multiple, spurious requests for the same data item, thereby making the particular item pseudo-popular. In other words, the item might not be popular (i.e., not requested by many clients), but the server is ignorant of this fact and considers it to be popular. The objective of hybrid scheduling is to remove this anomalous behavior and develop a performance analysis to obtain an estimate of average behavior of the real system. Intuitively, the spurious requests change the arrival rate in the pull system at every state. However, the service rate of both push and pull systems remains constant. Thus, the overall model of the system remains similar to the birth and death process as shown in Figure 5.10, but with different measures of λ_i and all $\mu_{2,i} = \mu_2$. Naturally, the state space and basic equations of the model is similar to equations (5.3–5.5). However, we need to estimate the different arrival rates at different states.

A careful look into this system reveals that the basic idea behind removal of the anomaly is to ignore multiple spurious requests for a data item sent by the same set of clients. While modelling and analyzing such a system is extremely complex, quite satisfactory results can be obtained by not considering the individual client's role explicitly. Hence, for performance analysis, we consider the system as ignoring the multiple, spurious requests for a particular data item as a whole. At this point in time we explain the behavior of the system characterized by the presence of data items. Note that every state in Figure 5.10 represents the number of items present in that state. Hence, in state $(1, 0)$ and $(1, 1)$ any one of the $D - K$ items would be present. Similarly, in state 2, any two items could be present, with the condition that an item already present (requested) will not be considered for another request. Thus, if item $K + 1th$ is already present, the possible items to be considered will be: $(K+2, K+3, \ldots, D)$. If item $K+2$ is present, other items could be $(K+1, K+3, \ldots, D)$, and so on. Hence, considering the states $(2, 0)$ and $(2, 1)$ in the system, we can say that the probability $Pr[uniq]_2$ that an item requested will not be requested again is given by:

$$Pr[uniq]_2 = \sum_{j=K+1}^{D} \rho_j \sum_{x=K+1, x \neq j}^{D} \rho_x$$

$$= 2! \sum_{j=K+1}^{D} \rho_j \sum_{x=j, x \neq j}^{D} \rho_{x+1}$$

$$\text{(as } \rho_j \rho_x = \rho_x \rho_j) \qquad (5.14)$$

This procedure goes on for all the following states. Thus, in every state we consider unique data items requested by clients. The probability that a requested item will not be requested again, is given by:

$$Pr[uniq]_i = i! \sum_{j=K+1}^{D} \rho_j \sum_{x=j}^{D-i+1} [\rho_{x+1} \cdots \rho_{x+i-1}]$$

$$= i! \sum_{j=K+1}^{D} \rho_j \sum_{x=j}^{D-i+1} \prod_{y=x+1}^{x+i-1} \rho_y \tag{5.15}$$

Hence, the arrival rate in the state that contains i items, is given by:

$$\hat{\lambda}_i = \lambda \sum_{j=K+1}^{D} \rho_j \sum_{x=j}^{D-i+1} \prod_{y=x+1}^{x+i-1} \rho_y$$

$$= \lambda \left(i! \sum_{j=1}^{n} \rho_j \sum_{k=j}^{n-i+1} [\rho_{x+1} \cdots \rho_{x+i-1}] \right)$$

$$(\text{as } \rho_j \rho_k = \rho_k \rho_j) \tag{5.16}$$

To obtain the probability $p(0, 0)$ of the idle state, we evaluate the expression at $z = 1$. Indeed, the occupancy of the push and pull states $\hat{P}_2(1) = \rho$ and $\hat{P}_1(1) = 1 - \rho$, where $\rho = \frac{\sum_{i=0}^{\infty} \lambda_i p(i,1)}{\mu_2}$. Thus, using suitable z transform of equations (5.4) and (5.5), we get:

$$\hat{P}_2(z) = \frac{1}{\mu_2}[z\hat{P}_1(z)(\hat{\lambda}_i + \mu_1)] - \frac{(\hat{\lambda}_i + \mu_1)zp(0,0)}{\mu_2}$$

$$+ p(1,1) - \frac{z^2 \hat{P}_1(z)\hat{\lambda}_{i-1}}{\mu_2} \tag{5.17}$$

To obtain the probability $p(0, 0)$ of the idle state, we evaluate the expression at $z = 1$. Indeed, the occupancy of the push and pull states is still the same. Thus, $\hat{P}_2(1) = \rho$ and $\hat{P}_1(1) = 1 - \rho$, where $\rho = \frac{\sum_{i=0}^{\infty} \lambda_i p(i,1)}{\mu_2}$. Thus we have:

$$\hat{P}_2(1) = \frac{(\hat{\lambda}_i + \mu_1)}{\mu_2}\hat{P}_1(1) - \frac{(\hat{\lambda}_i - \hat{\lambda}_0 + \mu_1)}{\mu_2}p(0,0) - \frac{\hat{P}_1(1)\hat{\lambda}_{i-1}}{\mu_2}$$

$$\rho = \frac{(\hat{\lambda}_i + \mu_1)}{\mu_2}(1 - \rho) - \frac{(\hat{\lambda}_i - \hat{\lambda}_0 + \mu_1)}{\mu_2}p(0,0) - \frac{\hat{\lambda}_{i-1}}{\mu_2}\hat{P}_1(1)$$

$$p(0,0) = \left[\frac{\hat{\lambda}_i + \mu_1}{\mu_2}(1 - \rho) - \rho - \frac{\hat{\lambda}_{i-1}}{\mu_2}(1 - \rho) \right] \left(\frac{\mu_2}{\hat{\lambda}_i - \hat{\lambda}_0 + \mu_1} \right) \tag{5.18}$$

where $\hat{\lambda}_i$ is given by equation (5.16). In order to get an estimate of the average system performance, we differentiate equation (5.17) to estimate the expected number of elements in the pull system.

$$\frac{d\hat{P}_2(Z)}{dZ} = \frac{\hat{\lambda}_i + \mu_1}{\mu_2}\left[\hat{P} - 1(Z) + Z\frac{d\hat{P}_1(Z)}{dZ}\right]$$

$$-\frac{\hat{\lambda}_i + \mu_1}{\mu_2}p(0,0) - \frac{2Z\hat{P}_1(Z)\hat{\lambda}_{i-1}}{\mu_2} - \frac{Z^2}{\mu_2}\frac{d\hat{P}_1(Z)}{dZ}\hat{\lambda}_{i-1}$$

$$\frac{d\hat{P}_2(Z)}{dZ}|_{Z=1} = \frac{\hat{\lambda}_i + \mu_1}{\mu_2}p(0,0) - \frac{2\hat{\lambda}_{i-1}}{\mu_2}\hat{P}_1(1) - \frac{\hat{\lambda}_{i-1}}{\mu_2}\frac{d\hat{P}_1}{dZ}|_{Z=1}$$

$$E_a[\mathscr{L}_{pull}] = \frac{\hat{\lambda}_i + \mu_1}{\mu_2}[1 - \rho - E[\mathscr{L}_{push}]] - \frac{\hat{\lambda}_i + \mu_1}{\mu_2}p(0,0)$$

$$-\frac{2\hat{\lambda}_{i-1}}{\mu_2}\rho - \frac{\hat{\lambda}_{i-1}}{\mu_2}\rho \qquad (5.19)$$

Subsequently, using Little's formulae and combining the expression for waiting time of push system, the expected access-time, $E_a[T_{hyb-acc}]$, of our hybrid system which considers anomalies, is obtained as:

$$E_a[T_{hyb-acc}] = \frac{K}{2(1-\rho)\mu_1}\sum_{i=1}^{K}P_i + E_a[W_{pull}] \times \sum_{i=K+1}^{D}P_i \qquad (5.20)$$

where $E_a[W_{pull}] = \frac{E_a[\mathscr{L}_{pull}]}{\lambda}]$.

5.3 SIMULATION EXPERIMENTS

In this section we validate the performance of our hybrid system through simulation experiments developed separately for both hybrid scheduling with client's departure and hybrid scheduling with anomalies. While the primary goal of hybrid scheduling with anomalies is to reduce the expected access time, the hybrid scheduling with client's departure also considers reducing the service drop, apart from minimizing the expected access time. Before presenting the details of simulation results, we enumerate the salient assumptions and parameters used in our simulation.

1. The simulation experiments are evaluated for a total number of $D = 1000$ data items.
2. The overall arrival rate λ' is varied between 1–4 arrivals per unit time. The value of μ_1 and μ_2 is estimated as: $\mu_1 = \sum_{i=1}^{K}(P_i \times L_i)$ and $\mu_2 = \sum_{i=K+1}^{D}(P_i \times L_i)$ where P_i and L_i are the access probability and length of data item i, respectively.
3. The lengths of data items are varied from 1 to 5.
4. To keep the access probabilities of the items from similar to very skewed, θ is dynamically varied from 0.20 to 1.40.
5. To compare the performance of our hybrid scheduling strategy with client's impatience, we have chosen the work in [24], as according to our knowledge, this is the only existing broadcast scheme which considered client's impatience.

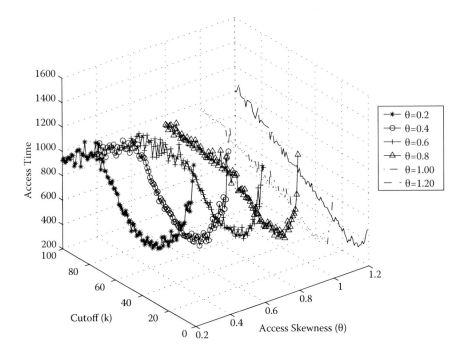

FIGURE 5.11 Expected access time with cutoff point.

In the following, we discuss a series of simulation results to demonstrate the efficiency of our two hybrid scheduling strategies. First we look into the results considering the client's departure (arising from impatience) from the system. Then we discuss the situation where client's impatience gives rise to anomalous behavior.

5.3.1 HYBRID SCHEDULING WITH CLIENT'S DEPARTURE

Figure 5.11 demonstrates the variation of expected access time with cutoff points (K) for different values of access skewness, θ. For all values of θ, with increasing K, the expected access time initially decreases to a certain point and then increases again. The reason is that with lower values of K, the access times for push items are pretty low while those for pull items are very high. The scenario is reversed when the value of K is pretty high. The curve for the expected access time takes a bell-shaped form, with the minimum value obtaining for a certain cutoff point, termed as optimal cutoff.

The different arrival rates of data items have significant impact on the minimum expected access time achieved by the system. Figure 5.12 shows that for different access skewness and with increasing arrival rates, the expected access time increases. For an arrival rate of 1 and 4, the average access time is in the range 100–400 and 400–750 time units, respectively.

Next we analyze the variation of the cutoff point with access skewness for different arrival rates. This is necessary to get a clear picture of the system dynamics, as the cutoff point plays the major role to minimize the expected access time. Figure 5.13

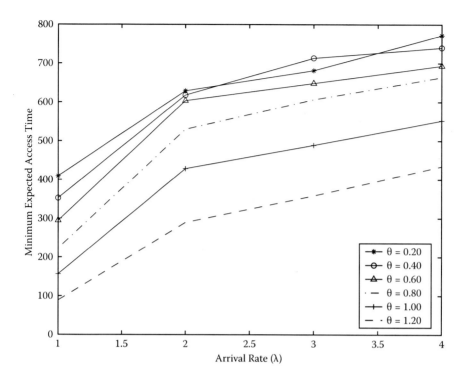

FIGURE 5.12 Minimum expected access time with arrival rates.

shows that the value of cutoff point decreases with increasing values of access skewness, θ. For example, $K = 300$–500 for lower skewness ($\theta \leq 0.6$) and $K = 100$–150 for higher skewness ($\theta \geq 1.00$). The reason is that with increasing skewness, the items get more skewed and number of popular items decreases. Hence, fewer number of items are pushed, decreasing the cutoff point.

One major objective of our proposed hybrid scheduling is to reduce the dropped requests arising from client's impatience. Figure 5.14 depicts the average number of requests dropped with access skewness for different arrival rates. The performance is compared with the existing strategy [24] for client's impatience in data broadcasting with unit arrival rate. As expected, the number of drop requests increases with increasing arrival rates. However, for all arrival rates, the number of drop requests is significantly lower than the number of drop requests in existing work. This is true even for higher arrival rates $\lambda' \geq 2$. This points out the efficiency of our hybrid scheduling strategy while considering client's departure due to impatience.

Figure 5.15 provides the comparative view between analytical and simulation results for hybrid scheduling with client's departure. The simulation results closely match the analytical results. The minor $\sim 8\%$ difference primarily is due to the analytical results only capturing an approximate average value.

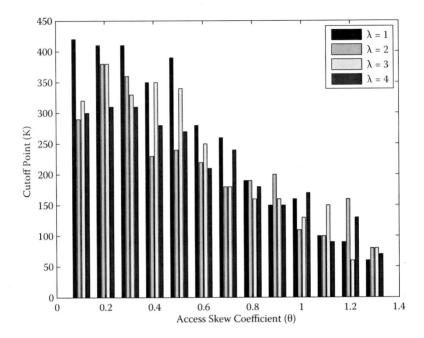

FIGURE 5.13 Variation of cutoff point.

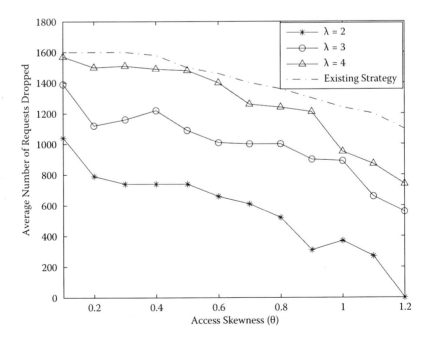

FIGURE 5.14 Average number of requests dropped.

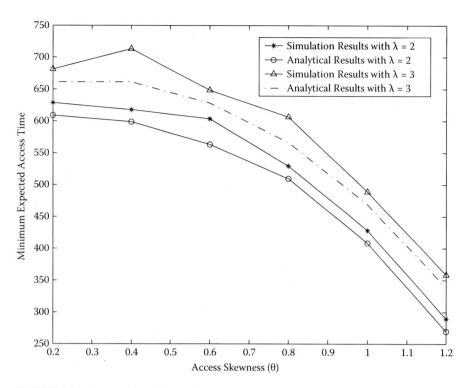

FIGURE 5.15 Comparison of analytical and simulation results.

5.3.2 HYBRID SCHEDULING WITH ANOMALIES

In this section, we discuss the simulation results for hybrid scheduling where the clients' impatience does not compel them to leave the system, but makes them transmit multiple requests for the same data item, thus generating an anomaly in the system.

Figure 5.16 delineates the variation of expected access time with cutoff point for different values of access skewness. The variation of this access time is similar to Figure 5.11, and takes a bell-shaped form, i.e., the expected access time first decreases up to a certain point and then starts increasing. The optimal value of K is chosen to get the minimum expected access time for hybrid scheduling with anomalies. The changes in the expected access time with different arrival rates is shown in Figure 5.17 for different values of access skewness. The increase in access skewness results in lower expected access time for all values of arrival rates. For items of unit length, the expected access time lies in the range 150–400 time units. For items of length 4, the expected access time is $\sim$ 50–110 time units. The change in minimum expected access time with different values of access skewness and item length is depicted in Figure 5.18. The waiting time is minimized (100 time units) for items of unit length and higher values of access skewness.

Finally, we investigate the dynamics of cutoff points with different access skewness and arrival rates. Figures 5.19 and 5.20 show the variation of cutoff point with access skewness for different values of arrival rates and item lengths, respectively.

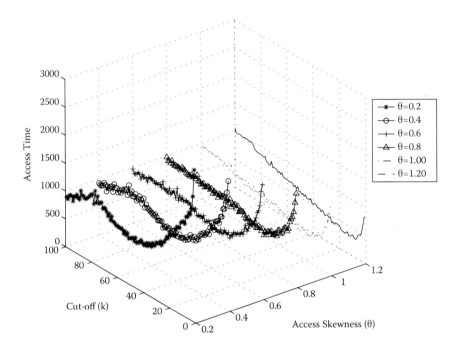

FIGURE 5.16 Variation of expected access time with cutoff point.

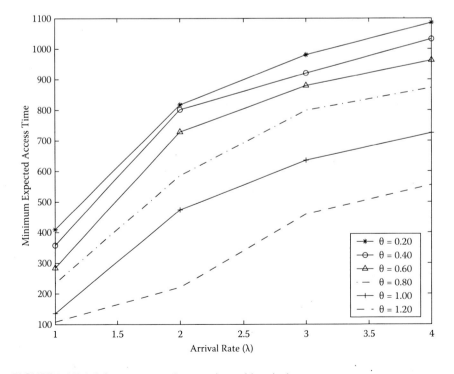

FIGURE 5.17 Minimum expected access time with arrival rates.

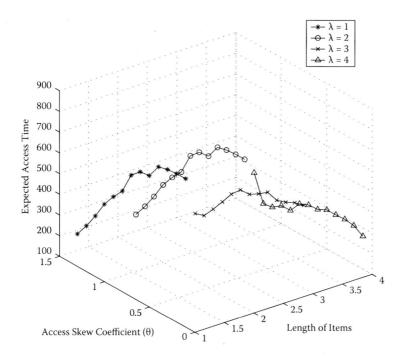

FIGURE 5.18 Expected access time with item length.

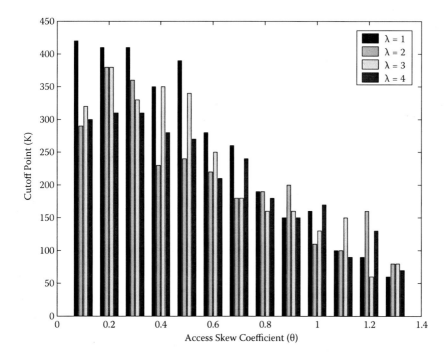

FIGURE 5.19 Variation of cutoff point with arrival rates.

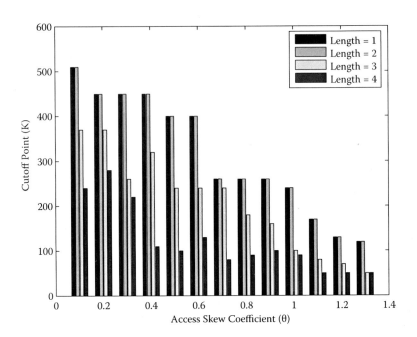

FIGURE 5.20 Variation of cutoff point with item length.

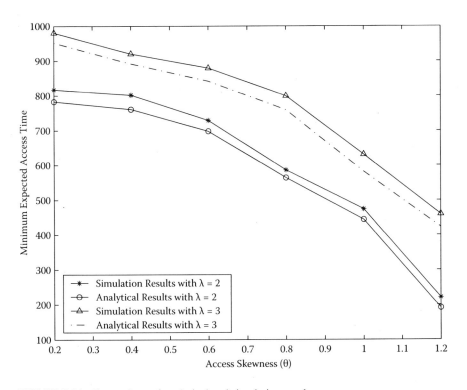

FIGURE 5.21 Comparison of analytical and simulation results.

For higher skewness, the cutoff decreases, thereby allowing more items in the pull queue and less items to be pushed. This is performed to achieve the minimum expected access time of the system.

Figure 5.21 provides the comparative view between analytical and simulation results in hybrid scheduling with anomalies. The simulation results closely match (90%) with the analytical results. The minor difference is again attributed to the approximate nature of the analysis.

5.4 SUMMARY

In this chapter we enhanced our hybrid scheduling framework to make it more practical and close to real systems. In real systems, the clients often become impatient which might result in two different scenarios. An impatient client might leave the system. Excessive impatience might lead to the client's declination to rejoin the system. On the other hand, an impatient client can send multiple requests for the same item (the item it wants), thereby increasing that item's popularity. The server (system) might be ignorant of this fact, and can consider the item as a popular one. This raises an anomaly in the system. In this chapter we have enhanced our hybrid scheduling framework to cope with clients' impatience to resolve the situation caused by clients' departure and spurious requests (anomalies).

6 Dynamic Hybrid Scheduling with Request Repetition

In this chapter we propose a dynamic hybrid scheduling [39], in which any new request for a pull item is kept in the pull queue. However, the clients' impatience resulting from prolonged waiting for any item, or a new request for the same data item by another client often makes them transmit repeated requests. The server keeps these repeated requests in the *repeat attempt (retrial)* queue, thereby distinguishing such requests from the new requests arriving in the pull queue. At any instant of time the item to be serviced is selected by using *stretch* (i.e, *max-request min-service-time first*) optimal scheduling algorithm. The service of an item from the pull queue needs to consider the service of the instances of same items from the repeat attempt queue also. Using a multidimensional Markov model, the average performance of the overall heterogenous, hybrid scheduling system is derived.

6.1 REPEAT ATTEMPT HYBRID SCHEDULING SCHEME

Most practical telecommunication systems often suffer from repeat attempt behavior from it's clients. As shown in Figure 6.1, particularly in the busy hour, the clients repeatedly request a particular data item. These repeated requests create a bottleneck in the server and wireless bandwidth that require special attention. Figure 6.2 is an overview of a repeat attempt system. In conventional communication, any request that finds the terminal busy is put on the waiting queue. In a repeat attempt model, however, a request that finds the server busy checks if the item is in the waiting queue. If not, the item is put in the waiting queue. If the item is already in the waiting queue, it is stored in the repeat attempt queue. This forms the basis of our newly proposed repeat attempt hybrid scheduling system. The database at the server consists of a total number of D distinct, heterogeneous items, from which K items are pushed and the remaining $(D - K)$ items are pulled. The access probability of an item (P_i), i.e., the popularity of the items among the clients, is governed by the Zipf's distribution and depends on the access skew coefficient (θ). From time to time the value of θ is changed dynamically for our hybrid system, thus varying P_i of all items and hence varying the size of the push and the pull sets dynamically.

The server maintains the database of all variable length items. Periodically the server pushes the data items using a broadcast schedule. We have used the packet fair scheduling (PFS) principle [19], that schedules the data items in an order such that two consecutive instances of the same data items are always equally spaced. When a client needs an item i, it sends its request for item i to the server and waits until it listens for i on the channel. If the request is for a push item, the server simply ignores

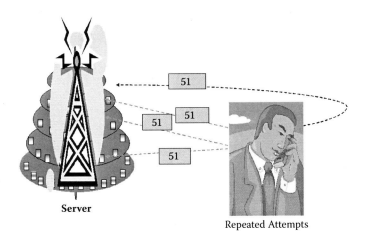

FIGURE 6.1 Repeated attempt from a mobile client.

the request as the item will be pushed according to the PFS algorithm. However, if the request is for a pull item, then the server first checks whether it is a new item request from a client or it is a request for the same data item by another client. If it is a request for a new item, it inserts the request into the pull queue with the arrival time and updates its stretch value. On the other hand, if the request is not a new one, i.e., some other client has already requested the item, the server considers it as a repeat attempt from an impatient client, inserts the item into the repeat attempt (retrial) queue and updates its stretch value. After every push, if the pull queue is not empty, the server chooses one item based on optimal stretch value, i.e., the item having *max-request min-service-time* value from the pull queue. It pulls that item and clears the pending requests for that item in the pull queue. Subsequently, the server now checks the repeat attempt queue and clears the requests associated with the instances of the same item. Figure 6.3 provides the pseudo-code of the repeat attempt, heterogeneous hybrid scheduling algorithm executing at the server side, where the procedure *Access and pull()* is depicted in Figure 6.4.

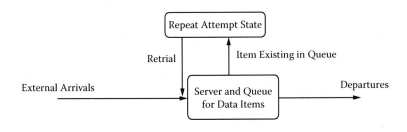

FIGURE 6.2 Overview of repeat-attempt system.

```
Procedure Hybrid Scheduling with Repeat Attempts;
while (true) do
begin
    Broadcast all the pages of an item, selected according to the PFS;
    Access and pull();
    if (the pull queue is not empty) then
      extract an item which optimizes the stretch from the pull queue;
      if (tie)
        extract the item with the smallest index;
      clear the number of pending requests for this item in the pull queue;
      clear the pending requests for the instance of the same item
      in the repeat attempt (retrial) queue;
      pull the particular item;
    Access and pull();
end;
```

FIGURE 6.3 Hybrid scheduling algorithm with repeat attempts.

6.2 PERFORMANCE ANALYSIS OF THE HYBRID REPEAT ATTEMPT SYSTEM

In normal pull-based scheduling strategy, the clients send explicit requests to the server and the server queues the requests. The item i with maximum requests or maximum stretch (number of requests/square of length i) is selected for service. However, in real systems the clients are often impatient, i.e., they often send multiple requests for a data item while it is not being serviced. Similarly, if a data item is already requested by a client and is waiting for service, and another client requests the same data item, the item is also considered as a repeat attempt item. In these scenarios, the data items having multiple requests are assumed to be in a new state, termed *repeat attempt state*.

We have assumed Poisson's arrival and exponential service of the items to make the analysis mathematically tractable. Figure 6.5 shows the schematic diagram of such a multidimensional Markov model representing the repeat attempt hybrid system.

```
Procedure Access and Pull();
while (true) do
begin
    take a specific number of accesses after broadcasting each page;
    if(the request is for pull item)
      if(new request)
        insert the request into the pull queue with arrival time;
      else
        mark the request as a repeat attempt;
        insert the request into the repeat attempt (retrial) queue;
end;
```

FIGURE 6.4 Access and pull scheduling.

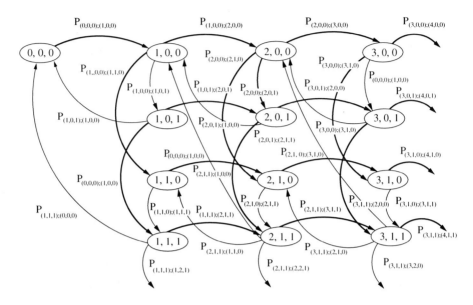

FIGURE 6.5 Repeat attempt Markov model of hybrid scheduling.

Any state of the system is represented by (x, y, z), where x represents number of unique items in the pull queue and y represents number of repeat attempt items in the repeat attempt queue and $z = 0$ (or 1) represents the push (or pull) system is currently under operation. The average arrival rate of the pull queue is assumed as λ. On the other hand, the arrival in the repeat attempt queue is assumed to be directly proportional of the number of items present in the pull queue. Thus, the arrival rate in the repeat attempt queue is taken as $xk\lambda$, where k is the scaling factor based on per item's average repeat attempt probability. We denote the transitional probability associated with transition from any state (x, y, z) to any another state (x', y', z') by $P_{(x,y,z);(x',y',z')}$. A careful insight into the system, shown in Figure 6.5 demonstrates the following major transitions:

1. Only a single transition is possible from initial (idle) state $(0, 0, 0)$. This happens with probability $P_{(0,0,0);(1,0,0)}$ during the arrival of any item in the pull system.
2. Arrival of any item in the pull queue results in transition of state in both the push and pull systems from $(x, y, 0)$ and $(x, y, 1)$ to $(x + 1, y, 0)$ and $(x + 1, y, 1)$ with probabilities $P_{(x,y,0);(x+1,y,0)}$ and $P_{(x,y,1);(x+1,y,1)}$, respectively.
3. Similarly, arrival of any item in the repeat attempt queue results in transition of states in the repeat attempt system from $(x, y, 0)$ and $(x, y, 1)$ to $(x, y + 1, 0)$ and $(x, y + 1, 1)$ with probabilities $P_{(x,y,0);(x,y+1,0)}$ and $P_{(x,y,1);(x,y+1,1)}$, respectively.
4. Service of an item in the push system results in transition of states from $(x, y, 0)$ to $(x, y, 1)$ with probability $P_{(x,y,0);(x,y,1)}$. However, depending on the number of repeated attempts, the service of an item in the pull system can result in transition of states from $(x, y, 1)$ to $(x - 1, y, 0)$,

$(x - 1, y - 1, 0), \ldots, (x - 1, 0, 0)$ with probabilities $P_{(x,y,1);(x-1,y,0)}$, $P_{(x,y,1);(x-1,y,0)}, \ldots, P_{(x,y,1);(x-1,0,0)}$, respectively. When the pull system contains only a single element, the service of an item results in transition from $(1, y, 1)$ to $(0, 0, 0)$ with probability $P_{(1,y,1);(0,0,0)}$.

To get the estimates of these transitional probabilities, first we need to derive the probabilities of selecting a particular item for service from the pull queue and repeat attempt queue. Subsequently, we need to obtain the relations between different service rates and measure for transition probabilities of the Markov chain.

For example, referring to the states $(2, 0, 0)$ (push with 2 items) and $(2, 0, 1)$ (pull with 2 items) in Figure 6.5, the arrival of a new pull item with arrival rate λ in the system, leads to the transition into state $(3, 0, 0)$ and $(3, 0, 1)$ with probability $P_{(2,0,0);(3,0,0)}$ and $P_{(2,0,0);(3,0,1)}$, respectively. Similarly, arrival of a repeat attempt item at these two states with an arrival rate $2k\lambda$ results in transition into the state $(2, 1, 0)$ and $(2, 1, 1)$ with probability $P_{(2,0,0);(2,1,0)}$ and $P_{(2,0,1);(2,1,1)}$, respectively. We have assumed strictly reciprocal service of a push and pull item. The average service rate of the push system is assumed to be μ'. Such a service of an item from the push system indicates that the next service will be from the pull system. Referring to the same state, i.e., $(2, 0, 0)$ in Figure 6.5, the service of the item results in transition from state $(2, 0, 0)$ to state $(2, 0, 1)$ with probability $P_{(2,0,0);(2,0,1)}$ and service rate μ'. However, the service of an item results in different possibilities, because the item currently getting serviced might be present or absent in the repeat attempt queue. If it is present in the repeat attempt queue, then the number of entries of that particular item in the repeat attempt queue also need to be cleared. Hence, service from state $(2, 1, 1)$ results in transition to either of the states $(1, 0, 0)$ or $(1, 1, 0)$ with probabilities $P_{(2,1,1);(1,0,0)}$ and $P_{(2,1,1);(1,1,0)}$ with service rates μ_1 and μ_2, respectively. We first need to discover the selection probabilities of different data items in the pull and repeat attempt queue. Because, there are x number of items currently present in the pull system, the actual items could be any combination of x elements chosen from total m data items in the system. Obviously, there are $\kappa = \binom{m}{x}$ number of combinations possible. We denote the combination by $\vec{\mathscr{C}} = \{\vec{C}_1, \vec{C}_2, \ldots, \vec{C}_\kappa\}$, where every $\vec{C}_i$ is an x element vector. Every element of this vector is a data item. We can select an element i from any of these vectors in $\binom{\kappa}{1}$ ways. Once we have chosen i from a particular vector every other item of the remaining $x - 1$ items can be chosen from any element of the available vectors. It should be noted that same items cannot be repeated, as repeated items reside in the *repeat attempt queue*. Moreover, since the pull queue does not contain the repeated instances of the items, the sum of total probability of the queue is less than 1. Hence all such probabilities need to be normalized. If p_i represents the access probability of item i, then normalized probability $Pr[i]_{norm}$ of choosing any item i is given by the relation:

$$Pr[i]_{norm} = \left(\frac{1}{\sum_{j=1}^{\kappa} Pr[\vec{C}_j]} \right) \binom{\kappa}{1} \left(p_i \sum_{j_1=1, j_1 \neq i}^{x} p_{j_1} \sum_{j_2=1, j_1 \neq i}^{x} p_{j_2} \cdots \sum_{j_\kappa=1, j_\kappa \neq i}^{x} p_{j_\kappa} \right)$$

(6.1)

where $Pr[\vec{C}_j]$ represents the probability of all the items belonging to the vector $\vec{C}_j$.

We now investigate the repeat attempt queue, where the elements can be repeated. They can be repeated once, twice, or up to m times. We are looking to obtain the probability of this repetition of elements. Proceeding in the similar approach as in equation (6.1), we can obtain the probability of a particular item i to be repeated any number of times in the repeat attempt queue. Let, $\left(Pr[i]_{Repeat}\right)_y$ denote the probability that the item i is repeated y times in the repeat attempt queue. Now, for the first time, the item i can still be selected in $\binom{\kappa}{i} = \kappa$ different ways. However, since i will be repeated once more, after choosing it once, it can still be selected in κ ways for the second time. The other terms for the remaining items can be chosen from any element of the available vectors. Proceeding in a similar way, the probability $\left(Pr[i]_{Repeat}\right)_y$ that there are y repetition of the item i is given by the equation:

$$\left(Pr[i]_{Repeat}\right)_y$$
$$= \kappa^y \left(p_i \sum_{j_1=1}^{x} \sum_{j_2=1}^{x} \cdots \sum_{j_{y-1}=1}^{x} \sum_{j_y=1, j_y \neq i}^{x} \cdots \sum_{j_\kappa=1, j_\kappa \neq i}^{x} p_{j_1} \cdots p_{j_{y-1}} p_{j_y} \cdots p_{j_\kappa} \right) \quad (6.2)$$

The normalized probabilities of repeat attempt states are now obtained by dividing the probability $\left(Pr[i]_{Repeat}\right)_y$ by the total probability of all the elements in the repeat attempt queue:

$$\left(Pr[i]_{Repeat}\right)_{y_{norm}} = \frac{\left(Pr[i]_{Repeat}\right)_y}{\sum_{i=1}^{x} \sum_{j=1}^{y} \left(Pr[i]_{Repeat}\right)_j} \quad (6.3)$$

It should be noted that when a departure occurs from a repeat attempt state, the next state always depends on the probabilities of the number of repeated attempts occurred. Thus, the fraction of overall service rate (μ) associated with no repetition and multiple (y) repetitions of the data item i is given by the following equations:

$$\mu_y = Pr[i]_{norm} \left(Pr[i]_{Repeat}\right)_{y_{norm}} \mu \quad \text{and}$$
$$\mu_0 = \left(Pr[i]_{norm} \left[1 - \left(Pr[i]_{Repeat}\right)_{1_{norm}} - \cdots - \left(Pr[i]_{Repeat}\right)_{y_{norm}} \right] \right) \mu \quad (6.4)$$

The expression for different transitional probabilities of the Markov chain is now given as:

$$P_{(x,y,0);(x+1,y,0)} = \frac{\lambda}{\lambda + xk\lambda + \mu'} \qquad P_{(x,y,0);(x,y+1,0)} = \frac{xk\lambda}{\lambda + xk\lambda + \mu'},$$

$$P_{(x,y,0);(x,y,1)} = \frac{\mu'}{\lambda + xk\lambda + \mu'} \qquad P_{(x,y,1);(x+1,y,1)} = \frac{\lambda}{\lambda + xk\lambda + \sum_{i=0}^{y} \mu_i},$$

$$P_{(x,y,1);(x,y+1,1)} = \frac{xk\lambda}{\lambda + xk\lambda + \sum_{i=0}^{y} \mu_i} \qquad P_{(1,y,1);(0,0,0)} = \frac{\mu_0}{\lambda + xk\lambda + \mu_0},$$

$$P_{(x,y,1);(x-1,y,0)} = \frac{\mu_0}{\lambda + xk\lambda + \mu_0} \qquad P_{(x,y,1);(x-1,y-1,0)} = \frac{\mu_1}{\lambda + xk\lambda + \mu_0},$$

$$\cdots\cdots \qquad \cdots\cdots\cdots$$

$$P_{(x,y,1);(x-1,0,0)} = \frac{\mu_y}{\lambda + xk\lambda + \mu_0}, \quad \forall y \geq 0 \quad P_{(0,0,0);(1,0,0)} = 1 \quad (6.5)$$

The transitional probabilities of the Markov chain obtained in this manner form the transitional matrix, containing the necessary information of the hybrid system. Any entry corresponding to (x, y, z), (x', y', z') in the transition matrix, actually contains the state transition probability $P_{(x,y,z);(x',y',z')}$ from (x, y, z) to (x', y', z'). Representing all the steady states by the vector $\vec{\pi}$ and the transition matrix by $\mathbf{P}$, an approximate measure of the steady state probabilities can be obtained by solving the following matrix equations associated with the Markov chain: $\vec{\pi} = \vec{\pi}\mathbf{P}$ and $\vec{\pi}\mathbf{e} = 1$, where $\mathbf{e}$ is a unit column vector. Solving the above equations helps us in obtaining the state probabilities $\pi = \{\pi(0, 0, 0), \ldots, \pi(x, y, z)\}$. The average number of items in the system and the average waiting time is now estimated as:

$$E[Items] = \sum_{x=0}^{m}\sum_{y=0}^{x} [\pi(x, y, 1) + \pi(x, y, 1)]$$
$$E[W] = E[Items]/\lambda \tag{6.6}$$

This provides an average behavior of our newly proposed hybrid scheduling system, which considers repeated attempts from the clients.

6.3 SIMULATION EXPERIMENTS

The simulation experiments are evaluated for a total number of $D = 1000$ data items of length between 1–10, average arrival rate $\lambda' \in [5, \ldots, 20]$ with access skew coefficient $\theta \in [0.20, \ldots, 1.40]$. To compare the performance of our hybrid system, we have chosen the hybrid scheduling strategy proposed in reference [32] as performance benchmarks. Figure 6.6 demonstrates the variation of the expected access time with

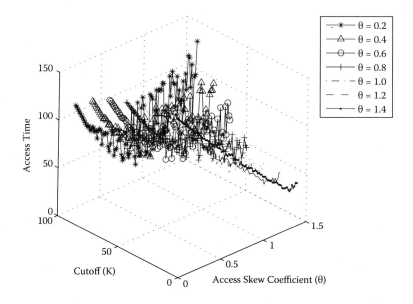

FIGURE 6.6 Performance of hybrid scheduling.

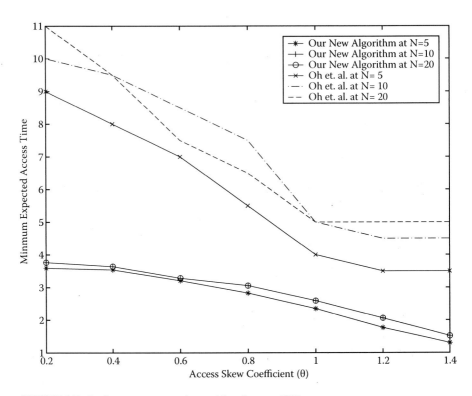

FIGURE 6.7 Performance comparison with reference [32].

different values of K and θ, for $\lambda = 10$, in our hybrid repeat attempt scheduling system. With increasing values of cutoff point K, the expected access time initially decreases, attains a minimum value, and then starts increasing again. This minimum point also provides the optimum cutoff point for which the framework gets an exact balance between the push and pull systems. Figure 6.7 shows the results of performance comparison, in terms of expected access time (in seconds), between our newly proposed repeat attempt hybrid framework with the existing hybrid scheme due to Oh, et al. [32]. The effective combination of PFS and stretch optimal scheduling strategies, together with the repeat attempt functionality results in the reduced waiting time in our hybrid scheduling framework. Figure 6.8 depicts the comparative view of the analytical results with the simulation results of our repeat-attempt hybrid scheduling framework. The analytical results closely match the simulation results for expected access time with almost $\sim 95\%$ accuracy, indicating that the performance analysis is capable of capturing the average system behavior with accuracy.

Figure 6.9 demonstrates the variation of cutoff point K for three different arrival rates $\lambda = [5, 10, 20]$. This indicates that the system achieved a fair balance between push and pull systems to attain the minimum expected access time.

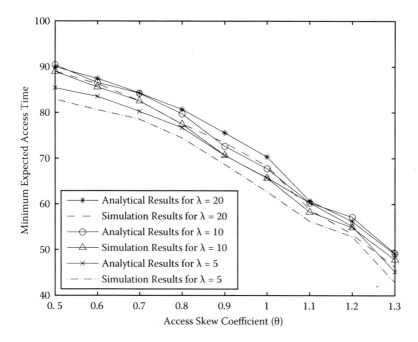

FIGURE 6.8 Simulation versus analytical results.

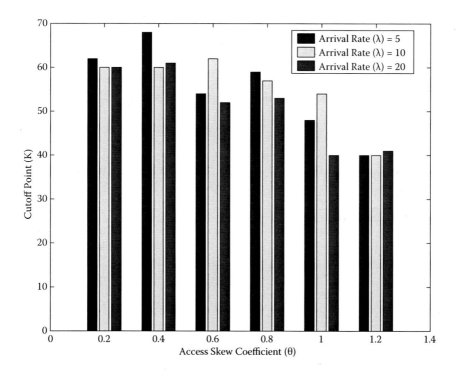

FIGURE 6.9 Variation of cutoff point (K).

6.4   SUMMARY

In this chapter we have enhanced our hybrid scheduling to incorporate the client's repeat attempt (retrial) behavior. The client's impatience often results in repeated attempts (retrials) for the same item. We have used suitable modeling, analysis, and simulation experiments to capture the clients' retrials.

7 Service Classification in Hybrid Scheduling for Differentiated QoS

In this chapter we propose a new service classification strategy [38], [42] for hybrid broadcasting to support the differentiated QoS in wireless data networks. The major novelty of our work lies in separating the clients into different classes and introducing the concept of a new selection criteria, termed as *importance factor* by combining the clients' priority and the stretch (i.e., max-request min-service-time) value. The item having the maximum importance factor is selected from the pull queue. The service providers now provide different service level agreements (SLA), by guaranteeing different levels of resource provisioning to each class of clients. Figure 7.1 shows one such service level agreement from the service provider by offering more bandwidth allocation for highest priority clients (e.g. Gold or Silver category of users); than for the lower priority clients basic users. The QoS (delay and blocking) guarantee for different classes of clients now becomes different, with the clients having maximum importance factor achieving the highest level of QoS guarantee. The performance of our heterogeneous hybrid scheduler is analyzed using suitable priority queues to derive the expected waiting time. The bandwidth of the wireless channels is distributed among the client classes to minimize the request blocking of highest priority clients. The cutoff point, used to segregate the push and pull items, is efficiently chosen such that the overall costs associated in the system are minimized.

7.1 HYBRID SCHEDULING WITH SERVICE CLASSIFICATION

The push-based broadcasting still ignores the clients' requests, and uses a *flat* round-robin scheduling strategy for cyclic broadcasting of popular data items. The pull scheduling, on the other hand, is based on a linear combination of the number of clients' requests accumulated and priorities. It should be noted that items with pending requests for higher priority clients should be serviced faster than the items having requests from lower priority clients. However, this scheme might suffer from unfairness to the lower priority clients and also does not consider the number of clients' requests. A data item, requested by many clients having lower importance, might remain in the pull queue for a long time. Eventually, all the pending requests for that item might be lost (blocked). Hence, a better option is to consider both the number of pending requests and the priorities of all clients requesting the particular data item. A close look into the system reveals that the service time required by an item is dependent on the size of that item. The larger the length of an item, the higher is its service time. We introduce a new scheduling strategy that combines stretch optimal or max-request min-service-time first schedule with the priority scheduling to select an item from the pull queue, shown in Figure 7.2.

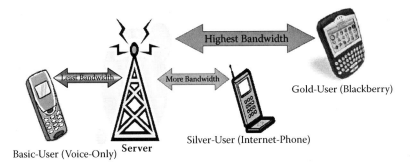

FIGURE 7.1 Differentiated allocation of wireless bandwidth.

Formally, if $\mathscr{S}_i$ represents the stretch associated with item i and $\mathscr{Q}_i$ represents the total clients' priority associated with item i, then the item selected from the pull queue is determined by the following condition:

$$\gamma_i = \max[\alpha \mathscr{S}_i + (1-\alpha)\mathscr{Q}_i] \tag{7.1}$$

where α is a fraction $0 \le \alpha \le 1$, that determines the relative weights between the priority and the stretch value. Clearly, $\alpha = 0$ and $\alpha = 1$ makes the schedule priority scheduling and stretch optimal scheduling, respectively.

When a client needs an item i, it requests the server for item i and waits until it listens for i on the channel. Note that the behavior of the client is independent of whether the requested item belongs to the push set or the pull set. Depending on the priorities, the server first classifies the clients into different service classes. The server goes on accumulating the set of requests from the clients. If the request is for a

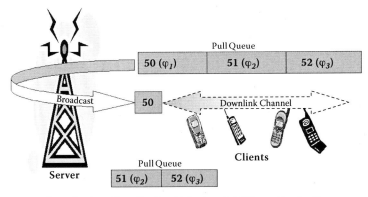

$\varphi_1 = \alpha$ (stretch value of item i) + (1 - a) (client-probabilities waiting for i)
$\varphi_1 = \alpha s_1 + (1 - a) p_k$ (for all k clients waiting for item i)
$s_1 = $ # of unique requests for item i / (item i's length)^2
$\alpha = $ any random number between 0 and 1

FIGURE 7.2 Hybrid priority scheduling for differentiated QoS.

```
Procedure HYBRID SCHEDULING;
divide the clients among different service classes;
while true do
    begin
    take out an item from the push scheduling and broadcast it;
    consider the access/requests arriving;
    ignore the requests for push item;
    append the requests for the pull item in the
    pull queue with its arrival time and importance factor;
    if the pull queue is not empty then
      extract the item having maximum importance factor
      (γᵢ) from the pull queue;
      if the required bandwidth for the item is
        greater than the available bandwidth for the
        corresponding service class then
        drop that item and the corresponding requests;
      else
          assign the required bandwidth of the item and
          update the available bandwidth;
          transmit that item;
          clear the number of pending requests for that item;
          free the amount of required bandwidth and update
          the amount of available bandwidth;
      take the set of access/requests arriving;
      ignore the requests for push item;
      append the requests for the pull item in
      the pull-queue with its arrival time and importance factor;
    end-if
end-while
```

FIGURE 7.3 Service classification in hybrid scheduling.

push item, the server simply ignores the request as the item will be pushed according to the online flat round-robin algorithm. However, if the request is for a pull item, the server inserts it into the pull queue with the arrival time, and updates the stretch value and total priority of all the clients requesting that item. After every push, if the pull queue is not empty, the server chooses the item having maximum importance factor (γ_i) from the pull queue. The bandwidth required by the data item is assumed to follow Poisson's distribution. If the required bandwidth of the data item is less than the bandwidth available for the corresponding service class, then the data item and the corresponding requests are lost. Otherwise, the server assigns the required bandwidth and transmits the item. Once the transmission is complete, the pending requests for that item in the pull queue is cleared and the bandwidth used is released to update the available bandwidth. Figure 7.3 provides the pseudo-code of the hybrid scheduling algorithm executing at the server side.

7.2   DELAY AND BLOCKING IN DIFFERENTIATED QoS

In this section we study the performance evaluation of our hybrid scheduler system by developing suitable models to analyze its behavior. The prime concern of this analysis is to obtain an estimate of the minimum expected waiting time (delay) of the hybrid system. Because this waiting time is dependent on the cutoff point K, investigation into the delay dynamics with different values of K is necessary to get the optimal cutoff point. As explained in Section 7.1, the selection criteria in the pull system is dependent on both the stretch value associated with the item and the priority of the clients requesting that particular item. Hence, the performance analysis also needs to consider the clients priority along with the stretch value associated with every data item. We divide the entire analysis into two parts. In the first part, we consider the system without any role of the client's priority and obtain the expression for the average number of items present in the system. In the second part, we introduce the explicit role of priorities in determining the average system performance.

7.2.1   AVERAGE NUMBER OF ELEMENTS IN THE SYSTEM

Assumptions: The arrival rate in the entire system is assumed to obey the Poisson's distribution with mean λ'. The service times of both the push and pull systems are exponentially distributed with mean μ_1 and μ_2, respectively. Let C, D, and K respectively represents maximum number of clients, total number of distinct data items, and the cutoff point. The server pushes K items and clients pull the rest $(D - K)$, of the items. Thus, the arrival rate in the pull system is given by: $\lambda = \sum_{i=K+1}^{D} \mathscr{P}_i \times \lambda'$, where $\mathscr{P}_i$ denotes the access probability of item i. We have assumed that the access probabilities P_i follow the *Zipf's distribution* with *access skew coefficient* θ, such that $\mathscr{P}_i = \frac{(1/i)^\theta}{\sum_{j=1}^{n}(1/j)^\theta}$.

Figure 7.4 illustrates the birth and death model of our system, where the arrival rate in the pull system is given by λ. Any state of the overall system is represented by the tuple (i, j), where i represents the number of items in the pull system and $j = 0$ (or 1) respectively represents whether the push system (or pull system) is being served. The arrival of a data item in the pull system results in the transition from state (i, j) to state $(i + 1, j)$, $\forall i \in [0, C]$ and $\forall j \in [0, 1]$. The service of an item

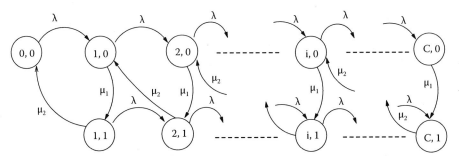

FIGURE 7.4 Performance modeling of hybrid system.

in the push system results in transition of the system from state $(i, j = 0)$ to state $(i, j = 1)$, $\forall i \in [0, C]$. On the other hand, the service of an item in the pull results in transition of the system from state $(i, j = 1)$ to the state $(i - 1, j = 0)$, $\forall i \in [1, C]$. The details of steady-state flow balance equations and their solutions are explained in Chapter 3 of the book. For the sake of clarity, we briefly highlight the major steps here. The steady-state behavior of the system (without considering priority) is represented by the equations given below:

$$p(0, 0) \lambda = p(1, 1) \mu_2$$
$$p(i, 0)(\lambda + \mu_1) = p(i - 1, 0)\lambda + p(i + 1, 1)\mu_2 \tag{7.2}$$
$$p(i, 1)(\lambda + \mu_2) = p(i, 0)\mu_1 + p(i - 1, 1)\lambda \tag{7.3}$$

where $p(i, j)$ represents the probability of state (i, j). Dividing both sides of equation (7.2) by μ_2, letting $\rho = \frac{\lambda}{\mu_2}$, $f = \frac{\mu_1}{\mu_2}$, performing subsequent z transform and using equation (7.2), we get

$$P_2(z) = \rho \, p\,(0, 0) + z\,(\rho + f\,)[P_1(z) - p\,(0, 0)] - \rho z^2\, P_1(z) \tag{7.4}$$
$$P_2(z) = \frac{f\,[P_1(z) - p(0, 0)]}{(1 + \rho - \rho z)} \tag{7.5}$$

Estimating the system behavior at the initial condition, we can state that the occupancy of pull and push states is given by: $P_2(1) = \sum_{i=1}^{C} p(i, 1) = \rho$ and $P_1(1) = \sum_{i=1}^{C} p(i, 0) = (1 - \rho)$. Using these two relations in equation (7.4), we can obtain the idle probability, $p(0, 0)$ as: $p(0, 0) = 1 - \rho - \frac{\rho}{f}$. Differentiating both sides of equation (7.4) with respect to z at $z = 1$, we estimate the expected number of elements in the pull system ($E[\mathcal{L}_{pull}]$) as follows:

$$\left[\frac{\partial P_2(z)}{\partial z}\right]_{z=1} = E[\mathcal{L}_{pull}]$$
$$= (\rho + f)\left[\frac{\partial P_1(z)}{\partial z}\right]_{z=1} + (1 - \rho) - (\rho + f) \times \left(1 - \rho - \frac{\rho}{f}\right) - \rho \mathcal{N} \tag{7.6}$$

where $\left[\frac{\partial P_2(z)}{\partial z}\right]_{z=1} = \mathcal{N}$ represents the average number of elements in the pull queue when a push request is being serviced.

7.2.2 PRIORITY-BASED SERVICE CLASSIFICATION

The analysis provided in Section 7.2.1 does not consider the priorities of the items residing in the pull queue. It only provides an estimation of the average number of items in the system, without considering the priorities. However, this analysis helps to develop the combined behavior of the pull system and its selection criteria. We now enumerate the basic assumptions used in our analysis:

1. It should be carefully noted that only the items in the pull subsystem are affected by the clients' priorities. The push system uses flat scheduling and is independent of the clients' priorities. To make the analysis tractable, we

assume that introducing the priorities in the pull system does not change the overall behavior of the push subsystem and does not affect the probability corresponding to the states in the push subsystem. This assumption leads us to the conclusion that the states of the upper chain in Figure 7.4 will not change with the inclusion of the clients' priorities. Only the pull states will be expanded reflecting the dynamics of items having different priorities.

2. Because, analytically it is not possible to get an estimate of the number of pending requests R_i for an item i, analytically we assume that the best possible estimate of this is the average number of instances of the particular item i present in the pull queue. Thus, $R_i = E[\mathscr{L}_{pull}]p_i$, and the stretch-value S_i is given by $S_i = \frac{R_i}{L_i^2}$.

3. Every client j is associated with a certain priority q_j, that reveals the importance or class of that client. If $E[\mathscr{L}_{pull}]$ represents the average length of the pull queue, then the average number of $i\,th$ items present in the queue is given by $E[\mathscr{L}_{pull}]\mathscr{P}_i$. Hence, the average importance of $i\,th$ item requested by $j\,th$ client is given by: $E[\mathscr{L}_{pull}]\,\mathscr{P}_i\,q_j$. Representing the influence of the set of clients $\mathscr{S}$ requesting for item i by $\mathscr{Q}_i = \sum_{j=1}^{\mathscr{S}} q_j$, the selection criteria of that element is now given by the following equation:

$$\rho_i = \left(\alpha \frac{E[\mathscr{L}_{pull}]\mathscr{P}_i}{L_i^2} + (1-\alpha)E[\mathscr{L}_{pull}]\,\mathscr{P}_i\,\mathscr{Q}_i \right) \tag{7.7}$$

This condition provides the position of every item in the priority queue. To distinguish this measure with the client priority q_j, we term ρ_i as the *importance factor* of item i. The items are grouped based on these importance factors. Formally, we can say if G_i represents the group of d_j items, then $\rho_{d_i} = \rho_j$. We analyze the system performance with clients belonging to two different classes [17], having two different importance factors.

7.2.2.1 Delay Estimation for Two Different Service Classes

Let, λ^1 and λ^2 represent the average arrival rate of the data items having importance factors 1 and 2, i.e., $\lambda = \lambda^1 + \lambda^2$. Formally we can say,

$$\lambda^i = \sum_{\forall j:\rho_j=\rho_i} \lambda_j \tag{7.8}$$

where $\lambda_j = \lambda\,\mathscr{P}_j$. We also assume that the most important items have the right to get service before the second important item without *preemption*. Now, the probability of every state should incorporate the number of items belonging to both important factors and the class of item currently getting service. We denote it by $p(m,n,r)$, such that: $p(m,n,r) = Pr[m$ and n units of importance factor 1 and 2 are present in the system and a unit of importance factor $r = 1(or\ 2)$ is in service in the pull system. Obviously, $r = 0$, when $m = n = 0$. Proceeding in a similar manner as shown in Section 7.2.1, we can obtain the steady state balanced equations of the prioritized pull system as:

$$(\lambda^1 + \lambda^2 + \mu_2)p(m, n, 2) = \lambda^1 p(m - 1, n, 2) + \lambda^2 p(m, n - 1, 2)$$
$$(\lambda^1 + \lambda^2 + \mu_2)p(m, n, 1) = \lambda^1 p(m - 1, n, 2) + \lambda^2 p(m, n - 1, 2) +$$
$$\mu_2[p(m + 1, n, 1) + p(m, n + 1, 1)]$$
$$(\lambda^1 + \lambda^2 + \mu_2)p(m, 1, 2) = \lambda^1 p(m - 1, 1, 2)$$
$$(\lambda^1 + \lambda^2 + \mu_2)p(1, n, 1) = \lambda^2 p(1, n - 1, 1) + \mu_2[p(2, n, 1) + p(1, n + 1, 2)]$$
$$(\lambda^1 + \lambda^2 + \mu_2)p(0, n, 2) = \lambda^2 p(0, n - 1, 2) + \mu_2[p(1, n, 1) + p(0, n + 1, 2)]$$
$$(\lambda^1 + \lambda^2 + \mu_2)p(m, 0, 1) = \lambda^1 p(m - 1, 0, 1) + \mu_2[p(m + 1, 0, 1) + p(m, 1, 2)]$$
$$(\lambda^1 + \lambda^2 + \mu_2)p(0, 1, 2) = \lambda^2 p(0, 0, 0) + \mu_2[p(1, 1, 1) + p(0, 2, 2)]$$
$$(\lambda^1 + \lambda^2 + \mu_2)p(1, 0, 1) = \lambda^1 p(0, 0, 0) + \mu_2[p(2, 0, 1) + p(1, 1, 2)] \qquad (7.9)$$

It should be noted that the probability of the idle state, i.e., $p(0, 0, 0) = p(0, 0)$ remains the same as before. The reason behind this is that the ordering of service does not affect the probability of idleness; i.e., $p(0, 0) = 1 - \rho - \frac{\rho}{f}$. Now, the occupancy of the pull states is ρ. Hence the fraction of time, the pull system is busy with type 1 and type 2 items is given by: $\rho\lambda^1/\lambda$ and $\rho\lambda^2/\lambda$. Thus we have,

$$\sum_{m=1}^{C}\sum_{n=0}^{C} p(m, n, 1) = \frac{\lambda^1}{\mu} \quad (a) \qquad\qquad \sum_{m=0}^{C}\sum_{n=1}^{C} p(m, n, 2) = \frac{\lambda^2}{\mu} \quad (b)$$
$$(7.10)$$

Obtaining a reasonable solution to these sets of stationary equations is almost impossible. All we can achieve is an expected measure of the system performance. Performing two successive z transforms over the equations 7.10 (a)–(b), we get,

$$P_{m1}(z) = \sum_{n=0}^{\infty} z^n p(m, n, 1) \qquad P_{m2}(z) = \sum_{n=1}^{\infty} z^n p(m, n, 2)$$

$$H_1(y, z) = \sum_{m=1}^{\infty} y^m P_{m1}(z) \qquad H_2(y, z) = \sum_{m=1}^{\infty} y^m P_{m2}(z) \qquad (7.11)$$

Combining the above two-dimensional z-transforms we have:

$$H(y, z) = H_1(y, z) + H_2(y, z) + p(0, 0, 0)$$
$$= \sum_{m=1}^{\infty}\sum_{n=1}^{\infty} y^m z^n [p(m, n, 1) + p(m, n, 2)] + \sum_{m=1}^{\infty} p(m, 0, 1)$$
$$+ \sum_{n=1}^{\infty} z^n p(0, n, 2) + p(0, 0, 0) \qquad (7.12)$$

Multiplying the set of steady state equations by suitable powers of y and z, summing up accordingly, and solving the above transforms result in:

$$H(y, z) = H_1(y, z) + H_2(y, z) + p(0, 0, 0)$$
$$= \frac{p(0, 0, 0)(1 - y)}{1 - y - \rho y(1 - z - \lambda^1 y/\lambda + \lambda^1 z/\lambda)}$$
$$+ \frac{(1 + \rho - \rho z + \lambda^1 z \mu_2)(z - y)P_{0,2}(z)}{\phi \times \psi}$$

where

$$\phi = z[1 + \rho - \lambda^1 y/\mu_2 - \lambda^2 z/\mu_2]$$
$$\psi = [1 - y - \rho y(1 - z - \lambda^1 y/\lambda + \lambda^1 z/\lambda] \qquad (7.13)$$

Now, if ζ_1 and ζ_2 represents the average number of items for both the classes, then

$$\zeta_1 = \left[\frac{\partial H(y,z)}{\partial y}\right]_{y=z=1} \qquad \zeta_2 = \left[\frac{\partial H(y,z)}{\partial z}\right]_{y=z=1} \qquad (7.14)$$

The expected waiting time of the data items having two different importance factors now can be found easily by using the Little's formula as: $E[W_1] = \zeta_1/\lambda^1$ and $E[W_2] = \zeta_2/\lambda^2$. Hence, the expected waiting time for clients with two different importance factors, in the hybrid system is:

$$E[T_{hyb\text{-}acc}]_A = \frac{1}{2\mu_1}\sum_{i=1}^{K} L_i \mathscr{P}_i + E[W_1]\sum_{i=k+1}^{D}\mathscr{P}_i \quad (a)$$

$$E[T_{hyb\text{-}acc}]_B = \frac{1}{2\mu_1}\sum_{i=1}^{K} L_i \mathscr{P}_i + E[W_2]\sum_{i=k+1}^{D}\mathscr{P}_i \quad (b) \qquad (7.15)$$

where, $E[W_2]$ is the (maximum) waiting time of the pull system, and where K is the cutoff point used to segregate push and pull components of the hybrid system. It should be noted that one major objective of our proposed algorithm is determining an optimal cutoff point K such that this delay is minimized.

7.2.2.2 Effect of Multiple Service Classes

The outline of the above procedure, however, fails to capture the expected system performance when the number of importance factors increase over 2. Thus, a better way is to follow a *direct expected value approach* [17]. Considering a nonpreemptive system with many importance factors, let us assume the data items with importance-factor ρ_j have an arrival rate and service time of λ^j and μ_{2j}, respectively. The occupancy arising due to this *jth* data item is represented by $\rho_j = \frac{\lambda^j}{\mu_{2j}}(1 \le j \le max)$, where *max* represents maximum possible value of importance factor. Also let σ_j represent the sum of all occupancy factors ρ_i, i.e., $\sigma_j = \sum_{i=1}^{j}\rho_i$. In the boundary conditions we have, $\sigma_0 = 0$ and $\sigma_{max} = \rho$. If we assume that a data item of importance factor i arrives at time t_0 and is serviced at time t_1, then the wait is $t_1 - t_0$. Let us assume at t_0 there are n_j data items present having priorities j. Also let, S_0 be the time required to finish the data item already in service, and S_j be the total time required to serve n_j. During the waiting time of any data item, n'_j new items having a higher importance factor can arrive and go to service before the current item. If S'_j is the total service time required to service all the n'_j items, then the expected waiting time for the *ith* item will be

$$E\left[W_{pull}^{(i)}\right] = \sum_{j=1}^{i-1}E[S'_j] + \sum_{j=1}^{i}E[S_j] + E[S_0] \qquad (7.16)$$

In order to get a reasonable estimate of $W_{pull}^{(i)}$, three components of equation 7.16 need to be individually evaluated.

(i) *Estimating $E[S_0]$*: The random variable S_0 actually represents the remaining time of service, and achieves a value 0 for idle system. Thus, the computation of $E[S_0]$ is performed in the following way:

$$
\begin{aligned}
E[S_0] &= Pr[\text{Busy-System}].E[S_0|\text{Busy-System}] \\
&= Pr[\text{Busy-Pull-Subsystem}].E[S_0|\text{Busy-Pull-Subsystem}] \\
&\quad + Pr[\text{Busy-Push-Subsystem}].E[S_0|\text{Busy-Push-Subsystem}] \\
&= \rho.\sum_{j=1}^{max} E[S_0|\text{Serving Pull-item with importance-factor} = j] \\
&\quad \times Pr[\text{Pull-item having importance-factor} = j] + (1 - \rho) \\
&\quad \times \sum_{j=1}^{max} E[S_0|\text{Serving Push-item with importance-factor} = j] \\
&\quad \times Pr[\text{Push-item having importance-factor} = j] \\
&= \rho \times \sum_{j=1}^{max} \frac{\rho_j}{\rho\mu_{2j}} + (1 - \rho) \times \sum_{j=1}^{max} \frac{\rho_j}{(1 - \rho)\mu_1} = \sum_{j=1}^{max} \frac{\rho_j}{\mu_{2j} + \mu_1}
\end{aligned}
$$

$$(7.17)$$

(ii) *Estimating $E[S_j]$*: The inherent independence of Poisson's process gives the flexibility to assume the service time $S_j^{(n)}$ of all n_j customers to be independent. Thus, an estimate of $E[S_j]$ can be obtained using the following steps:

$$
\begin{aligned}
E[S_j] &= E\left[n_j S_j^{(n)}\right] = E[n_j]E\left[S_j^{(n)}\right] \\
&= E[n_j]E\left[S_j^{(n)}\right]_{pull} + E[n_j]E\left[S_j^{(n)}\right]_{push} \\
&= \frac{E[n_j]}{\mu_{2j} + \mu_1}
\end{aligned}
$$

$$(7.18)$$

(iii) *Estimating $E[S'_j]$*: Proceeding in a similar way and assuming the uniform property of Poisson's,

$$
E[S'_j] = \frac{E[n'_j]}{\mu_{2j} + \mu_1}
$$

$$(7.19)$$

The solution of the above equations can be achieved by using Cobham's iterative induction [17] . The expected waiting time of the ith item and the overall expected waiting time of the pull system are given as:

$$
E\left[W_{pull}^{(i)}\right] = \frac{\sum_{j=1}^{max} \rho_j/(\mu_{2j} + \mu_1)}{(1 - \sigma_{i-1})(1 - \sigma_i)}
$$

$$
E\left[W_{pull}^q\right] = \sum_{i=1}^{max} \frac{\lambda^i E\left[W_{pull}^{q(i)}\right]}{\lambda}
$$

$$(7.20)$$

The overall expected access time is obtained by combining the time taken to service the push and pull items. Because, the push set contains K items of heterogeneous lengths $L_1, L_2, \ldots, L_K$, the average length of the push (broadcast) cycle is $\frac{1}{2} \sum_{i=1}^{K} L_i \mathscr{P}_i$. Thus, the expected access time ($E[T_{hyb\text{-}acc}]$) for different client classes and the overall average expected access time of our hybrid system are now given by:

$$E[T_{hyb\text{-}acc}]_j = \frac{1}{2\mu_1} \sum_{i=1}^{K} L_i \mathscr{P}_i + E\left[W_{pull}^{(j)}\right] \sum_{i=k+1}^{D} \mathscr{P}_i \quad (a)$$

$$E[T_{hyb\text{-}acc}] = \frac{1}{2\mu_1} \sum_{i=1}^{K} L_i \mathscr{P}_i + E\left[W_{pull}^{q}\right] \sum_{i=k+1}^{D} \mathscr{P}_i \quad (b) \qquad (7.21)$$

where K is the cutoff point used to segregate push and pull components of the hybrid system. It should be noted that one major objective of our proposed algorithm is to determine an optimal cutoff point K such that this delay is minimized. The above expression provides an estimate of the average delay (waiting time) for different classes of clients in our hybrid scheduling system. The service providers always try to reduce the delay of the high priority clients, to ensure their satisfaction. Apart from this delay, we would like an estimate of the prioritized cost associated with each class of client. This cost is actually obtained as $q_j \times E[T_{hyb\text{-}acc}]$. Intuitively, this cost provides an estimate of the client's influence on the service provider and the overall system.

7.3   SIMULATION EXPERIMENTS

In this section we validate the performance analysis of our prioritized hybrid system by performing simulation experiments. Because the framework is made for differentiated services in wireless data networks, the primary objective is to reduce the cost associated in maintaining the different classes of clients, thereby reducing the loss that might incur from the churning of the clients. Naturally, the high priority class of clients should encounter the minimum possible waiting time, because the system suffers more losing these highest priority clients. We first enumerate the set of assumptions used in our simulation. Subsequently, we provide the series of simulation results obtained.

7.3.1   ASSUMPTIONS

1. The simulation experiments are evaluated for a total number of data items $D = 100$.
2. The overall average arrival rate λ' is assumed to be 5. The values of μ_1 and μ_2 are estimated as: $\mu_1 = \sum_{i=1}^{K}(\mathscr{P}_i \times L_i)$ and $\mu_2 = \sum_{i=K+1}^{D}(\mathscr{P}_i \times L_i)$.
3. The lengths of the data items are varied from 1 to 5, with an average of 2.
4. To keep the access probabilities of the items from similar to very skewed, θ is dynamically varied from 0.20 to 1.40. More specifically, we have assumed $\theta = \{0.20, 0.60, 1.0, 1.40\}$.
5. The entire set of clients is divided into three classes: *Class A, having highest priority, Class B with medium priority*, and *Class C with lowest priority*.

The priorities are taken in the ratio 1 :: 2 :: 3. The fraction α associated in deriving the importance factor is assumed to be in the range [0, 1], where $\alpha = 1$ indicates the system ignoring the effect of priority and $\alpha = 0$ indicates the system ignoring the effect of stretch. The simulation experiments are performed for $\alpha = \{0, 0.25, 0.50, 0.75, 1.0\}$.

6. The distribution of clients among different classes is also assumed to obey Zipf's distribution, with the lowest number of highest priority (Class A) clients and the highest number of lowest priority clients.

7. The cost associated in maintaining the three different classes of clients is assumed to be inversely proportional to the priority of the classes. In other words, the costs associated with Class A, Class B, and Class C clients are assumed to be in the ratio 3 :: 2 :: 1.

7.3.2 RESULTS

The goal of the first set of experiments is to investigate the overall delay experienced by each class of clients. Figures 7.5–7.9 demonstrate the dynamics of total delay with the cutoff point experienced by the three different classes for $\alpha = \{0, 0.25, 0.50, 0.75, 1.0\}$, respectively. This is performed for different values of access skewness. The delay associated with the Class A (highest priority) clients is very low (within 5–10 broadcast units). The delay experienced by the Class B clients remains in the range 20–40 broadcast units. The highest delay (40–70 broadcast units) is experienced by the Class C clients. However, for all the classes of clients the delay is higher for low values of cutoff point (K). This is because for low values of K, the system deviates from the hybrid nature and cannot achieve an optimum balance between push and pull set.

The major objective of the second set of experiments is looking into the variation of the prioritized cost associated with each client class θ. As previously mentioned, the system assigns the cost to each class of clients in proportion to the priority of that particular class. Figure 7.10 demonstrates the variation of prioritized costs with the cutoff point, associated with each class of clients for $\alpha = \{0.25, 0.75\}$ and $\theta = 0.60$. The overall objective is to pick up the particular value of cutoff point to minimize the total prioritized cost.

Figure 7.11, on the other hand, shows the changes in total optimal prioritized cost of all the client classes, with different values of α for $\theta = \{0.20, 0.60, 1.40\}$. With decreasing values of α, the influence of priority increases and the prioritized cost reduces. The underlying reason is that for lower values of α, the increased influence of priority results in serving the important clients first, thereby reducing the overall cost of the system.

Figure 7.12 indicates the changes in the cutoff point with different values of θ for $\alpha = \{0, 0.75, 1.0\}$. For small values of θ, the cutoff point lies in the range 40–55. This is because for small θ the items have similar probabilities and the system obtains a very good balance between the push and pull set. However, for higher values of θ the probabilities of the items become skewed. Thus, the size of the push set (having high probabilities) shrinks, resulting in low cutoff points (in the range 15–20).

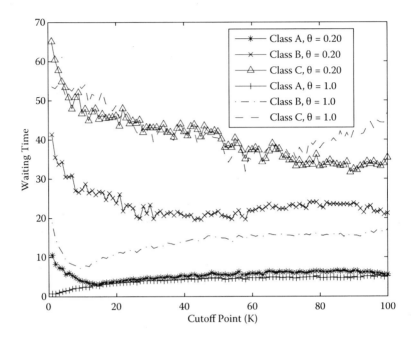

FIGURE 7.5 Delay variation with $\alpha = 0.0$.

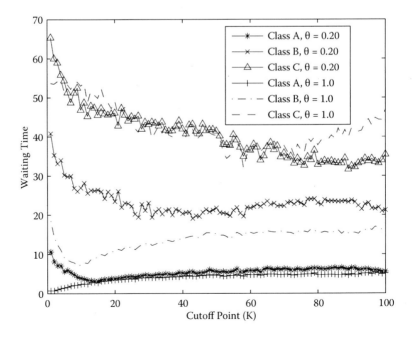

FIGURE 7.6 Delay variation with $\alpha = 0.25$.

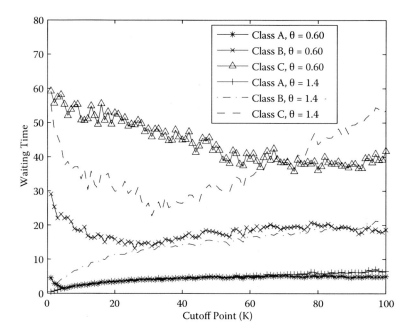

FIGURE 7.7 Delay variation with $\alpha = 0.50$.

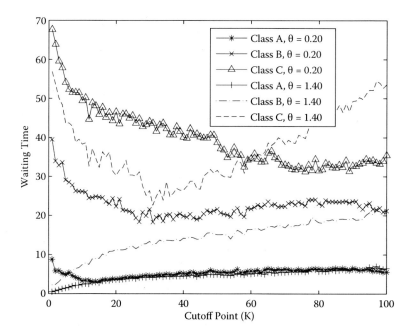

FIGURE 7.8 Delay variation with $\alpha = 0.75$.

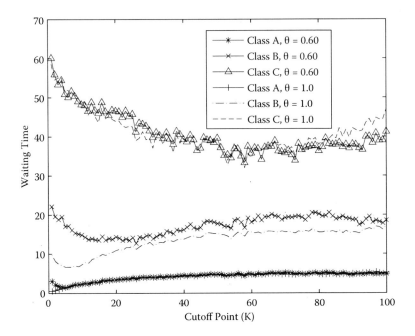

FIGURE 7.9 Delay variation with $\alpha = 1.0$.

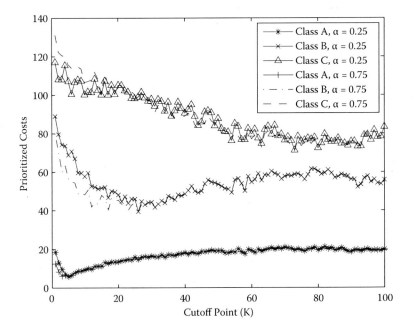

FIGURE 7.10 Cost dynamics for service classes.

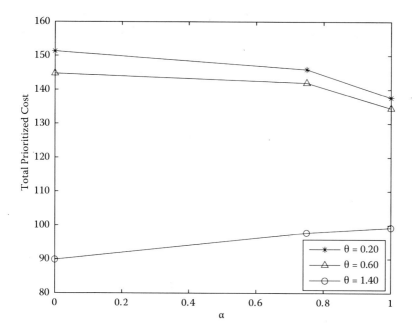

FIGURE 7.11 Variation of prioritized cost.

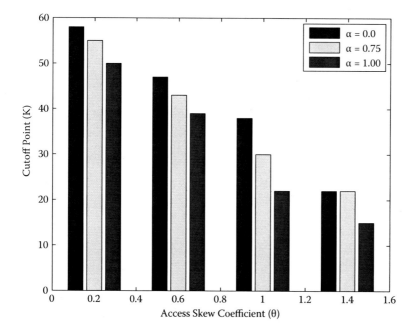

FIGURE 7.12 Variation of cutoff point.

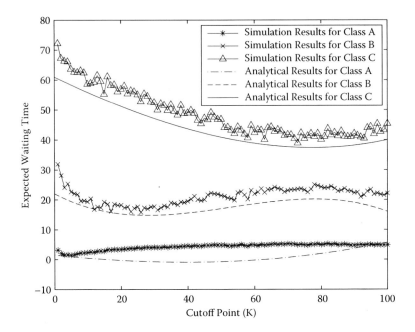

FIGURE 7.13 Analytical versus simulation with $\theta = 0.60$, $\alpha = 0.75$.

Figure 7.13 demonstrates the comparison between analytical and simulation results for $\theta = 0.60$ and $\alpha = 0.75$. The analytical results are obtained using the equation 7.21(a). We have chosen the values of α and θ so that these values are almost in the middle of their range. Analytical results closely match simulation results for all three sets of clients, with a minor 10% deviation. The minor deviation is attributed to the memoryless assumption in the system modelling.

7.4 SUMMARY

In this chapter we proposed a new priority-based service classification scheme suitable for differentiated QoS. Subsequently, we enhanced our hybrid-scheduling strategy by using this service classification scheme. The scheme explores clients' priorities and items' popularity for differential distribution of wireless resources. This results in a lower churning rate, improved QoS, and more profit for the service providers.

8 Online Hybrid Scheduling over Multiple Channels

This chapter investigates a new online hybrid solution [40] for the multiple broadcast problem. The new strategy first partitions the data items among multiple channels in a balanced way. Then, a hybrid push-pull schedule is adopted for each single channel, as shown in Figure 8.1.

Clients may request desired data through the uplink and go listen to the channel where the data will be transmitted. In each channel, the push and pull sets are served in an interleaved way: One unit of time is dedicated to an item belonging to the push set and one to an item of the pull set, if there are pending client requests not yet served. The push set is served according to a flat schedule, while the pull set according to the most requested first policy. Complete knowledge is not required in advance of the entire data set or of the demand probabilities, and the schedule is designed online.

Many real systems exist based on multiple-channels technology. A real multiple channel server in Figure 8.2 is shown while, a multiple channel broadcast is illustrated in Figure 8.3.

8.1 PRELIMINARIES: DEFINITIONS AND METRICS

Let $D = \{1, 2, \ldots, N\}$ be a set of N data items of unit length, and let each item i be characterized by a demand probability P_i. To begin, consider a system with a single broadcast channel. A *broadcast schedule* S of any period is an ordered sequence of data items selected from the set D. Note that if S is cyclic, then the period is a positive integer; otherwise, *period* $\to \infty$. Position t of S indicates the item of D that is broadcast at time $\tau \equiv t \bmod period$. The same item can be replicated in S. The *average spacing* between two consecutive instances of the same item i in S is termed s_i. Note that if i appears only once in S, then $s_i = period$. For total push systems, the expected item delay t_i for item i on S is defined as the average time a client waits before receiving i, assuming that, at any instant of time, clients start to listen with the same probability. Hence, $t_i = \frac{s_i}{2}$, for $1 \le i \le N$. Thus, the average expected delay (AED) is given by:

$$AED(D) = \sum_{i=1}^{N} t_i P_i = \frac{1}{2} \sum_{i=1}^{N} s_i P_i \qquad (8.1)$$

is the average over all items of D of their delay.

For the total pull systems, let $\delta_{i,r}$ be the actual delay between the request r for item i and the transmission time of item i. If R_i and $\#_i$ denote the set of requests for item i and its size, then the average item delay is defined as

$$\frac{\sum_{r \in R_i} \delta_{i,r}}{\#_i}$$

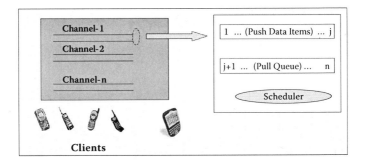

FIGURE 8.1 Dynamic scheduling over multiple channels.

FIGURE 8.2 Example of a multiple channel server.

FIGURE 8.3 Multiple channel broadcast.

and the average access time (AAT)

$$AAT(D) = \sum_{i=1}^{N} \frac{\sum_{r \in R_i} \delta_{i,r}}{\#_i} P_i \qquad (8.2)$$

is the average over all items of D of their average item delay.

For the hybrid push-pull systems, let $D = \Pi \cup \Delta$, where Π and Δ are the push and pull sets, respectively. Then, their performance is measured as the hybrid time (HAT), represented by:

$$HT(D) = AED(\Pi) + AAT(\Delta) \qquad (8.3)$$

Finally, the single broadcast problem is defined as the problem of finding the broadcast schedule S which minimizes equations 8.1, 8.2, and 8.3 for push, pull and hybrid systems, respectively.

Note that, for total push schedules, particular assumptions lead to simplified formulations. For example, for a flat schedule F, since $s_i = N$ for $1 \leq i \leq N$ and $\sum_{i=1}^{N} P_i = 1$, it holds that

$$AED_F(D) = \frac{N}{2}$$

For a schedule generated by the square root rule algorithm SRR, if the optimal spacing $s_i = (\sum_{j=1}^{N} \sqrt{P_j}) \sqrt{1/P_i}$ can be guaranteed, $AED_{SRR}(D) = (\sum_{j=1}^{N} \sqrt{P_j})^2$. Because in general, however, optimal spacing is not reachable when conflicts arise on the same schedule position, the following weaker result holds [48]

$$AED_{SRR}(D) \geq \left(\sum_{j=1}^{N} \sqrt{P_j} \right)^2$$

Consider now a system with K broadcast channels. Clearly, a *multiple broadcast schedule M* consists of K single broadcast schedules, one per channel. For total push systems, the average delay t_i for item i is defined exactly as in the single channel environment, except that two item occurrences are considered consecutive if they happen to be close in time, irrespective of which channels they appear on. Specifically, for a client listening simultaneously to the first j channels, two occurrences of i are consecutive, and they are s_i apart if an occurrence of item i appears at time τ_i on channel j_1, the subsequent earliest occurrence of i appears at time τ_{i+s_i} on channel j_2, with $1 \leq j_1 \leq j_2 \leq j$, and no other occurrence appears in any other channel between 1 and j at the instant of time $\tau_{i+1}, \ldots, \tau_{i+s_i-1}$. Now, let $AED^j(D)$ denote the AED experienced by a client listening to the first j channels. It is clear that a lower bound for $AED^j(D)$ is $\frac{AED(D)}{j}$. Note that, such a lower bound holds either when all data items are transmitted on each channel or when only a group of the data items is transmitted on each channel. Finally, the multiple average expected delay (*MAED*) is defined as the AED averaged over all the subsets of channels that clients can afford to read. To simplify, let clients listen only to consecutive subsets of channels, starting from channel 1. Thus, if a client listens to j channels, with $j > 1$, it will listen to channels $1, 2, \ldots, j$. Denoting by π_j the probability that clients listen to j channels, and assuming that $\sum_{j=1}^{K} \pi_j = 1$,

$$MAED(D) = \sum_{j=1}^{K} AED^j(D)\pi_j \qquad (8.4)$$

Clearly, MAED = AED when $K = 1$.

As for AED, also simplified expressions of MAED hold. Namely, for the multiple schedule based on the SRR [48], we have,

$$MAED_{SRR}(D) \geq \sum_{j=1}^{K} \frac{\frac{1}{2}\left(\sum i = 1^N \sqrt{P_i}\right)^2}{j}\pi_j \qquad (8.5)$$

Moreover, MAED boils down to a much simpler expression when skew allocation among channels and flat schedules SF are assumed [12, 53]. Indeed, assume that the data items are partitioned into K groups $G_1, G_2, \ldots, G_K$, where the group G_j consists of the N_j data items transmitted by a flat schedule on channel j. Since each item is transmitted only by a channel, MAED is bounded by a constant only if clients listen to all channels. Hence, assuming $\pi_K = 1$ and $\pi_j = 0$, for any $1 \leq j \leq K - 1$, it is easy to see that, for any skewed allocation,

$$MAED_{skew}(D) = AED^K(D) = AED_F(G_1) + \cdots + AED_F(G_K)$$

$$= \frac{1}{2}\sum_{j=1}^{K}\left(N_j \sum_{i \in G_j} P_i\right) \qquad (8.6)$$

Hence, the allocation problem, proposed in references [12, 53], consists of finding the skewed allocation and flat schedule in such a way that equation 8.6 is minimized. Note at this point that such SF schedule, (denoted from now on as SF), can be found by a dynamic programming strategy in $O(NK \log N)$ time, as shown in [12].

Nonetheless, since the allocation problem is a special case of the multiple broadcast problem, it is not known how far the optimal MAED of the allocation problem is from the optimal solution of the multiple broadcast problem, even when it is considered restricted to the push systems. In conclusion, although AAT and HT performance measures can be generalized to the case of multiple channels, we are not aware of solutions already proposed in literature for the multiple broadcast problem for total pull or hybrid systems.

8.2   A NEW MULTICHANNEL HYBRID SCHEDULING

The previous discussion suggests that many different schedules for the multiple broadcast problem can be obtained by combining different data allocation strategies with different schedule strategies for single channels. The solution proposed in this chapter for N data items and K channels combines a balanced allocation of data among channels with hybrid push-pull schedule per each single channel. The hybrid push-pull strategy guarantees that our solution adapts easily to changes of item demand-probability, while the balanced data allocation provides an easy way to incorporate new data items without any data preprocessing. Moreover, since no more than $\lceil N/K \rceil$ items are assigned to each channel, by the flat schedule, MAED cannot go beyond $\lceil \frac{N}{2K} \rceil$. Finally, because each client knows in advance the channel on which the desired item will be transmitted, it can listen only to a channel per time.

First, the *balanced K channel allocation with flat schedule*, briefly *BF*, solution is presented in Subsection 8.2.1. The performance of this simple solution is competitive with the *MAED* of both the *SRR* and *SF* schedules when all the items have almost the same demand probabilities. Then, in subsection 8.2.2, the flat schedule is substituted by the hybrid schedule to make our solution competitive even when the item demand probabilities are skewed.

8.2.1 Balanced K Channel Allocation with Flat Broadcast Per Channel

The balanced data allocation strategy, which assigns $O(N/K)$ items to each channel, lies in the opposite end of the skewed allocation strategy adopted for the K allocation Problem [53]. Specifically, consider a set of N data items $D = \{1, \ldots, N\}$ and K channels, numbered from 1 to K. The items are partitioned in K groups $G_1, \ldots, G_K$, where group $G_j = \{i | (i-1) \bmod K = j - 1\}$, whose size

$$N_j = \begin{cases} \lceil \frac{N}{K} \rceil & \text{if } 1 \le j \le (N \bmod K) \\ \lfloor \frac{N}{K} \rfloor & \text{if } (N \bmod K) + 1 \le j \le K \end{cases}$$

The balanced data allocation algorithm is depicted in Figure 8.4. The items assigned to each channel are then broadcast locally by a flat schedule. Specifically, item i assigned to group G_j will be broadcast as the $\lceil i/K \rceil$ *th* item of the flat schedule of channel j. Thus, the MAED of the *BF* schedule is given by the following relation:

$$MAED_{BF}(D) = \frac{1}{2} \sum_{j=1}^{N \bmod K} \left(\left\lceil \frac{N}{K} \right\rceil \sum_{i \in G_j} P_i \right) + \frac{1}{2} \sum_{j=N \bmod K + 1}^{K} \left(\left\lfloor \frac{N}{K} \right\rfloor \sum_{i \in G_j} P_i \right) \quad (8.7)$$

It can be seen that $\lfloor \frac{N}{2K} \rfloor \le MAED_{BF}(D) \le \lceil \frac{N}{2K} \rceil$. *BF* is periodic and independent of the demand probabilities. Moreover, it is easy to see how the *BF* schedule can be updated when the size N of the set of data items increases by one. More specifically, the new item $N + 1$ will become the $\lceil (N+1)/K \rceil$ *th* item of channel $j = (N \bmod K) + 1$.

> **Algorithm balanced K channel allocation :**
> **begin**
> for $i = 1, \ldots, N$ do
> $j = ((i-1) \bmod K) + 1;$
> $G_j = G_j \cup \{i\}$
> **end**

FIGURE 8.4 The balanced allocation algorithm.

TABLE 8.1
Performance Comparison

Algorithm	$N;\theta$ 2500; 0	$N;\theta$ 2500; 0.1	$N;\theta$ 2500; 0.2	$N;\theta$ 2500; 0.4	$N;\theta$ 10; 0.8	$N;\theta$ 500; 0.8
SRR	312.5	311.65	308.75	294.201	1.13	45.20
SF	312.5	311.86	309.69	298.44	1.17	47.53
BF	312.5	312.5	312.5	312.5	1.25	62.5

Table 8.1 compares the performances of the *BF* schedule, the *SF* schedule, and the lower bound, of the performance of the *SRR* schedule, as given in equation 8.5. The demand probabilities of the items are assumed to follow the Zipf distribution whose skew coefficient is θ; i.e., $P_i = \frac{(1/i)^{\theta}}{\sum_{i=1}^{N}(1/i)^{\theta}}$, for $1 \leq i \leq N$. The parameters N and θ have been chosen to range, respectively, in $10 \leq N \leq 2500$ and $0 \leq \theta \leq 0.8$, while K is fixed to 4. The demand probabilities become skewed as θ approaches 1. Evaluating the distance in percentage between the *MAED* of the *BF* schedule and the *MAED* of the *SF* schedule as

$$\varepsilon = \frac{MAED_{BF} - MAED_{SF}}{MAED_{SF}}$$

it is clear that the distance between the simple *BF* schedule is no larger that 4% for $\theta \leq 0.4$, leading to a very satisfying tradeoff between efficiency and simplicity. However, the gap is marked for large values of θ. Our new hybrid scheduling per channel, proposed in the following section, improves in this respect.

8.2.2 ONLINE BALANCED K CHANNEL ALLOCATION WITH HYBRID BROADCAST PER CHANNEL

In this section, we investigate the improvement on the MAED of the *BF*-algorithm, while using the balanced K channel allocation to partition the data items among the channels, but by replacing the flat schedule with the following new hybrid schedule at each channel. Figure 8.5 explains this multichannel, asymmetric, hybrid communication environment. For each channel j, the hybrid algorithm first partitions the group G_j assigned to each channel in two sets: the *push set* Π_j, whose items $\{1, \ldots, k_j\}$ will be broadcast according to a flat schedule, and the *pull set* Δ_j, whose items will be sent on demand. The hybrid schedule alternates between the transmission of one item extracted from the push set and the transmission of one item from the pull set. At each pull turn, the item to be sent on demand is the item most requested so far by clients. Note that the push set may gain several consecutive turns if there are no pending requests for items of the pull set. The algorithm that runs at the client site is depicted in Figure 8.6. A client, desiring to receive item i, sends to the server an explicit request through the uplink if $i > k_j$. Then, it goes to listen to channel $j = (i - 1) \mod K + 1$ to which item i has been assigned by the balanced allocation algorithm, and waits until i is transmitted.

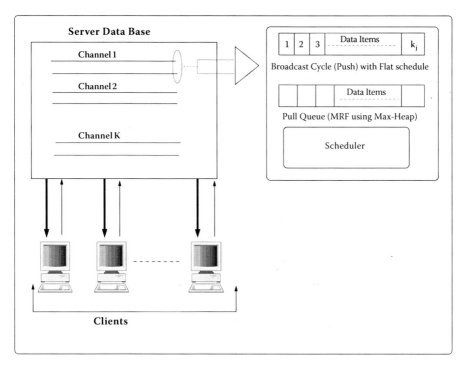

FIGURE 8.5 Multichannel hybrid scheduling environment.

The algorithm at the server site is illustrated in Figure 8.7. For each channel j, the server stores in F_j the flat schedule of Π_j, whose current length is k_j. For each item i of the pull set, the server maintains the number $\#_i$ of requests received between two consecutive transmissions of that item in a max-heap H_j. The requests are checked before each push turn. Note that only items of the pull set can be requested. The item broadcast at the pull turn is the one stored in the heap root, that is the item that has received so far the largest number of requests. After the pull transmission of item i, $\#_i$ is always set to 0. Note that to choose the next item to be pushed costs constant time, while $O(\log \Delta_j)$ is required to maintain the heap after each delete

```
Algorithm Client-Request (item i):
/* i : the desired item */
begin
j = (i − 1) mod k + 1;
if i > kⱼ then send to the server the
request for item i;
wait on channel j until i is transmit-
ted;
end
```

FIGURE 8.6 The client request algorithm at the client side.

Algorithm Hybrid (channel j, pull set Π_j, push set Δ_j);
while (true) **do**
begin
    check the requests received after the last check;
    for every item i that has been requested **do**
        $\#_i = \#_i + 1$;
        update H_j;
    broadcast the current item of F_j;
    update F_j;
    if $(H_j \neq \emptyset)$ **then**
        $i = root(H_j)$;
        pull item i;
        if $\#_i > \sigma$ **then move** i **from** Π_j **to** Δ_j;
        $\#_i = 0$;
end;

FIGURE 8.7 The hybrid algorithm at the server side.

max operation. The new algorithm is online since it decides at run time the new item to be transmitted. To have a schedule adaptive to noticeable changes of the demand probabilities, a mechanism for dynamically varying the push and pull sets is given based on the threshold σ. When item i is broadcast at the pull turn, if $\#_i > \sigma$, i is inserted in the push set as the last item of the flat schedule. Observe that although the push set initially consists of consecutive items, it may become fragmented. Then, the client needs more information to learn to which set the desired item belongs. More precisely, the server will supply an index of the changes occurring at the push sets, which is sent in a compressed form along with each single data item, and periodically defragmentation policies are applied to globally renumber the data items.

Still to be discussed the *MAED* performance of the balanced K channel allocation with hybrid schedule per channel, briefly *BH*, algorithm. As for the *BF* algorithm, the performance of *BH* is bounded by a constant only if the clients listen to all channels. Hence, $\pi_k = 1$ and $\pi_j = 0$ for $1 \leq j \leq k-1$. Restricted to the push sets, *BH* reduces to a *BF* schedule. Recalling that clients must afford to listen to all channels, its *MAED* performance measure is given by:

$$MAED_{BH}(D) = \gamma MAED_{SF}(\Pi_1 \cup \Pi_2 \cup \ldots \Pi_k) + \sum_{j=1}^{K} AAT(\Delta_j)$$

$$= \gamma \sum_{j=1}^{K} AED_F(\Pi_j) + \sum_{j=1}^{K} AAT(\Delta_j)$$

$$= \gamma \sum_{j=1}^{K} \sum_{i=1}^{k_j} \frac{k_j}{2} P_i + \sum_{j=1}^{K} \sum_{i=k_j+1}^{N_j} \frac{\sum_{r \in R_i} \delta_{r,i}}{\#_i} P_i \qquad (8.8)$$

where γ is the *interleaving coefficient* and varies from 1, when only push turn occur, to 2, when every push turn is followed by a pull turn. When the push sets are small, the time spent at the *BF* schedule becomes shorter, but the pull sets are larger, leading to longer access time if the system is highly loaded. Thus, the two sets should be chosen in such a way that they reflect the load of the system to gain the advantages of both push and pull based schedules.

8.3 SIMULATION RESULTS

In this section results of simulation experiments are used to discuss the performance of our multichannel scheduling strategies. First, a summary of the major assumptions and parameters used for our experiments.

1. The simulation experiments are evaluated for a total number of data items $N = 2000$.
2. The request arrival time is assumed to obey Poisson distribution with mean $\lambda = 10$, which simulates a middle load of the system. The item requests follow the Zipf's distribution, defined in Section 8.2.1. The average service time for every request is assumed to be 1.
3. The number of channels varies between 2–4.
4. The demand probabilities follow the Zipf's distribution with θ dynamically varied from 0.30 to 1.30.

Finally, note that all the experiments involving *BH* and *SRR* schedules were executed ten times, and the performance average reported.

8.3.1 RESULTS

Figure 8.8 shows the performance of the *SF* and *BF* schedules for $N = 2000$, $K = 3$, and $0.3 \leq \theta \leq 1.3$. Note that the results are independent of the arrival time of the requests, because two total push schedules are considered. Clearly, the *MAED* performance of the *BF* schedule is also independent from the demand probabilities, as shown in equation 8.7. The *SF* schedule results in significant gains in overall expected access time for high values of θ. The major objective of our *BH* schedule is to reduce such a gap performance.

Figure 8.9 demonstrates the performance efficiency of the *BH* schedule over the *SF* schedule. In addition to the *BH* schedule, as described in Section 8.2.2, a variant that maintains in each channel the item of the push set sorted by decreasing demand probabilities is studied. Note that the ordering is local to each channel. In this way, the push set is initialized with hot items, and similarly the pull set with the cold items. So from now on, let *BH-random* denote the basic hybrid schedule, while *BH decreasing* the new variant. Both the *BH* schedules result in an improvement of almost half of

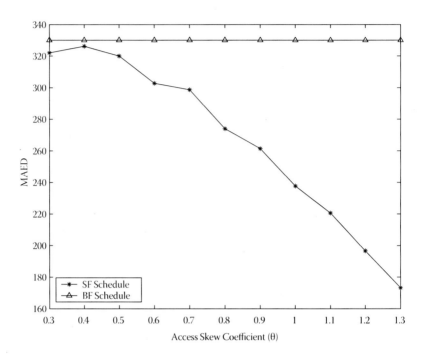

FIGURE 8.8 *MAED* performances of the *BF* and *SF* schedules.

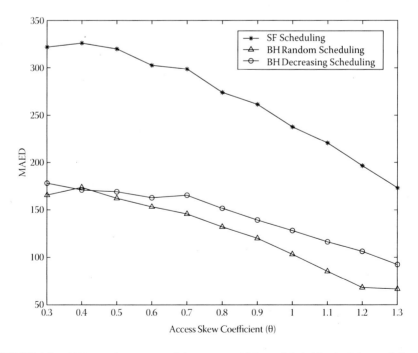

FIGURE 8.9 *MAED* performances of the new multichannel hybrid schedules and the *SF* schedule.

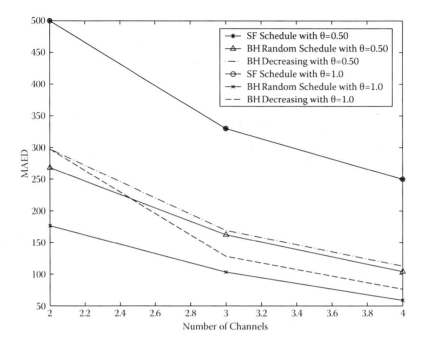

FIGURE 8.10 *MAED* performances of the *SF*, *BH* random and *BH* decreasing schedules.

the performance of the *SF* schedule. Figure 8.10 shows the gain achieved by *BH* schedules on different numbers of channels, for $\theta = 0.50$ or $\theta = 1.00$. Even for different number of channels, the *BH* schedules achieve almost half of the *MAED* measure in comparison to the *SF* schedule.

We have also taken a look into the distribution of items among different channels. While the balanced allocation distributes the items equally among the channels and the skewed allocation is twisted, different sizes of the push sets of the *BH* schedule can be selected initially. Figures 8.11 and 8.12 show that, for a fixed value of the skew parameter θ of the Zipf distribution, and assuming all the push sets initially empty, when the *BH* schedule reaches a steady state (i.e., when the threshold mechanism does not move any items), the sizes of the push sets are skewed at least as much as that of the groups of the *SF* schedule when $\theta \le 0.4$ and much more skewed for larger values of θ. Finally, Figure 8.13 shows that both *BH* random and *BH* decreasing schedules outperform the *SRR* schedule. Note that the *SRR* schedule is determined online, but as for all push systems, it only knows the system load through the demand probabilities. So, for fixed θ, while *BH* reacts to the changes of the load of the system because it listen to the clients, *SRR* cannot.

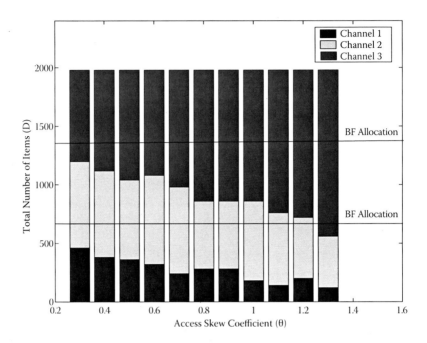

FIGURE 8.11 Size of the channel groups of the *SF* schedule when $K = 3$.

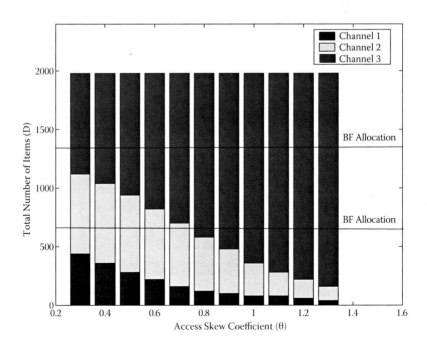

FIGURE 8.12 Size of the push sets of the *BH* schedule when $K = 3$.

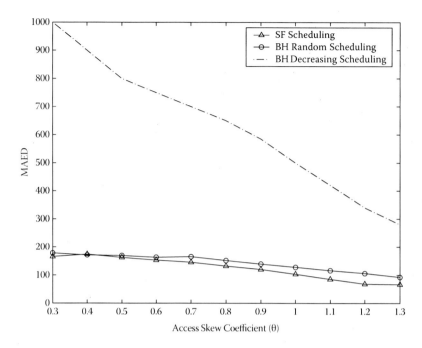

FIGURE 8.13 Performance comparison of the *SRR* and *BH* schedules.

8.4 SUMMARY

In this chapter we enhanced our hybrid scheduling strategy to span multiple channels. The data items are partitioned in an online, round-robin fashion over all the channels. Dynamic hybrid scheduling is then applied over every channel. The scheme significantly gains over the existing optimal skewed partition, followed by push scheduling in each channel.

9 Conclusions and Future Works

We have discussed a detailed framework for hybrid scheduling in asymmetric wireless environments. The scheduling and data transmission strategies can be broadly classified into push and pull scheduling schemes. However, both of these push and pull scheduling strategies suffer from some specific disadvantages. Hence, a hybrid scheduling that explores the advantages of both push and pull scheduling is more attractive. We first develop a basic hybrid scheduling scheme that combines the push and pull scheduling schemes independent of the *build-up point*, i.e., without restricting the pull queue size to be 1. Our hybrid scheduling system, uses push scheduling to broadcast the popular data items and takes the help of pull scheduling to transmit the less popular ones. The system computes the packet fair scheduling (PFS) for push system and accumulates the clients' request in the pull queue. The pull system works on most request first (MRF) scheduling. The system alternatively performs one push and one pull method. The cutoff point, that segregates between the push and pull scheduling is chosen in such a manner that the overall average access time experienced by clients is minimized. This hybrid scheduling strategy is enhanced to incorporate heterogeneous data items (i.e., items of variable lengths). Although the basics of the push schedule remain unchanged, pull scheduling now must also consider the item lengths. The underlying reason is that items of variable lengths have different service times. This leads us to use the stretch optimal scheduling principle to choose an item from the pull queue. Performance analysis and simulation results point out the efficiency of this heterogeneous hybrid scheduling scheme.

While this scheduling strictly alternates between one push and one pull method, a better approach is to adapt the operations depending on the overall system load. Hence, we further improve the hybrid push-pull scheduling to introduce multiple consecutive push and pull operations depending on the overall load and dynamism of the system. A procedure for providing the performance guarantee of the system to meet the deadline specified by the clients was also developed. A close look at the practical systems reveals that in most systems some clients might be impatient. This impatience significantly affects the performance of the system. An impatient client can leave the system even before the request is actually serviced. Excessive impatience might result in clients antipathy in joining the system again. On the other hand, an impatient client might send multiple requests for the same data item, thereby increasing the access probability of the item. This develops an anomalous picture of the system, because the server might consider the item very popular, which is not the actual case. The effects of such spurious requests from impatient clients must be resolved. We developed a hybrid scheduling principle to take care of clients' impatience and resolves the anomalous behavior. Performance modelling using birth and death processes and *multi dimensional Markov chain* is provided to capture an overall estimate of such practical hybrid scheduling principles.

Today's wireless PCS networks classify the clients into different categories based on importance. The goal of the service providers is to provide the highest priority clients with maximum possible satisfaction, even at the cost of some lower priority clients. This, in turn, helps to reduce the overall churn rate and increase the overall profit of the service providers. The role of clients' priorities must be considered to implement such *differentiated QoS*. We have introduced a new service classification scheme in our hybrid scheduling strategy that combines the stretch optimal and priority-based scheduling in a linear fashion to develop a new selection criteria, termed importance factor. While the items to be pushed are determined using a flat scheduling, the item from the pull queue is selected based on this importance factor. Modelling and analysis of the system is performed to get an average behavior of the QoS parameters like delay, bandwidth, and drop request in this new hybrid scheduling framework. The dissertation then investigates the hybrid scheduling over multiple channels. It shows that using online partitioning of data items into multiple channels and deploying hybrid schedule on every channel has the power to improve the average waiting time of the clients over the existing optimal multichannel push-based scheduling schemes.

While the wireless communication technology is rapidly enhancing from a voice-alone framework to an audio-visual world of real time applications, the efficiency of data broadcasting needs to be improved significantly. The gradual deployment of $3G$ wireless systems and the quest for $4G$ wireless systems has encouraged us to look at and investigate different current problems in data broadcasting. In the future, we want to look into the effects of efficient data caching mechanisms to save the energy of power constrained mobile devices. The dynamism of the wireless networks and the Internet often creates uncertainty and variation of the QoS parameters. We believe that the QoS offered in wireless networks and the Internet should not be constant, but needs to change over the time. Thus, to provide the services with some level of QoS guarantee, the QoS parameters should be renegotiated at specific intervals. We would like to look into the effects and solutions required to design such renegotiable QoS in data broadcasting over wireless systems and the Internet.

References

1. S. Acharya, R. Alonso, M. Franklin and S. Zdonik, "Broadcast Disks: Data Management for Asymmetric Communication Environmens," *Proc. of ACM SIGMOD*, 1995.
2. S. Acharya, M. Franklin and S. Zdonik, "Balancing Push and Pull for Data Broadcast," *Proc. of ACM SIGMOD*, 1997.
3. S. Acharya and S. Muthukrishnan, "Scheduling On-demand Broadcasts: New Metrics and Algoritms," *Proc. ACM Intl. Conf on Mobile Computing and Networking (MobiCom)*, 1998.
4. D. Aksoy and M. Franklin, "Scheduling for Large-Scale On-Demand Data Broadcasting," *Proc. of IEEE InfoCom*, 1998.
5. D. Aksoy and M. Franklin, "RxW: A Scheduling Approach for Large-Scale On-Demand Data Broadcast," *IEEE/ACM Trans. on Networking*, Vol. 7, No. 6, pp. 646–860, 1999.
6. D. Aksoy, M. Franklin and S. Zdonik, "Data Staging for On-Demand Broadcast," *Proc. of 27th VLDB Conference*, 2001.
7. A. Bar-Noy and Y. Shilo, "Optimal Broadcasting for Two Files over an Asymmetric Channel," *Journal of Parallel and Distributed Computing*, Vol. 60, No. 4, pp. 474–493, 2000.
8. A. Bar-Noy, J. S. Naor and B. Schieber. Pushing Dependent Data in Clients-Providers-Servers Systems. In *Mobile Networks and Applications*, Vol. 9, pp. 421–430, 2003.
9. A. Bar-Noy, B. Patt-Shamir and I. Ziper, "Broadcast Disks with Polynomial Cost Functions," *ACM/Kluwer Wireless Networks* (WINET), Vol. 10, pp. 157–168, 2004.
10. A. Bar-Noy, B. Patt-Shamir and I. Ziper, "Broadcast Disks with Polynomial Cost Functions," *ACM/Kluwer Wireless Networks* (WINET), Vol. 10, pp. 157–168, 2004.
11. Z. Brakereski and B. Patt-Shamir, "Jitter-Approximation Tradeoff for Periodic Scheduling," *Proc. of IEEE Intl. Workshop of WMAN*, 2004.
12. A. A. Bertossi, M. C. Pinotti, S. Ramaprasad, R. Rizzi and M. V. S. Shashanka, "Optimal Multi-channel Data Allocation with Flat Broadcast Per Channel," *Proc. of IEEE IPDPS*, 2004.
13. S. E. Czerwinski, B. Y. Zhao, T. D. Hodes, A. D. Joseph and R. H. Katz, "An Architecture for a Secure Service Discovery," *Proc. 5th Int. Conf. Mobile Computing (Mobicom)*, pp. 24–35, Aug. 1999.
14. M. Franklin and S. Zdonik, "A Framework for Scalable Dissemination-Based Systems," *Proceedings of the 12th ACM SIGPLAN conference on object-oriented programming, systems, languages, and applications*, pp. 94–105, 1997.
15. Q. Fang, V. Vrbsky, Y. Dang and W. Ni, "A Pull-Based Broadcast Algorithm that Considers Timing Constraints," *Proc. of Intl. Workshop on Mobile and Wireless Networking*, 2004.
16. J. Fernandez and K. Ramamritham, "Adaptive Dissemination of Data in Time-Critical Asymmetric Communication Environments," *Euromicro Conf. on Real-time Systems (ECRTS)*, 1999.
17. D. Gross and C. M. Harris, Fundamentals of Queuing Theory, *John Wiley & Sons Inc.*
18. Y. Guo, S. K. Das and M. C. Pinotti, "A New Hybrid Broadcast Scheduling Algorithm for Asymmetric Communication Systems: Push and Pull Data Based on Optimal Cut-Off Point," *Mobile Computing and Communications Review (MC2R)*, Vol. 5, No. 4, 2001.

19. S. Hameed and N. H. Vaidya, "Efficient algorithms for scheduling data broadcast," *Wireless Networks*, Vol. 5, pp. 183–193, 1999.

20. W-C. Lee, Q. Hu and D. L. Lee, "Channel Allocation Methods for Data Dissemination in Mobile Computing Environments," *Proc. of Intl. Symp. on High Performance Distributed Computing (HPDC)*, 1997.

21. Q. Hu, D. L. Lee and W-C. Lee, "Performance Evaluation of a Wireless Hierarchical Data Dissemination Systems," *Proc. of AGM MobiCom*, 1999.

22. C-L. Hu and M-S. Chen, "Adaptive Information Dissemination: An Extended Wireless Data Broadcasting Scheme with Loan-Based Feedback Control," *IEEE Trans. on Mobile Computing*, Vol. 2, No. 4, 2003.

23. http://www.direcpc.com, 1997.

24. S. Jiang and N. H. Vaidya, "Scheduling Data Broadcast to Impatient Users," *ACM Intl. Workshop on Mobile Data Engineering*, 1999.

25. C. Kenyon, N. Schanbanel and N. Young, "Polynomial-Time Approximation Scheme for Data Broadcst," *Proc. of ACM Symp. on Theory of Computing (STOC)*, 2000.

26. S. Khanna and V. Liberatore, "On Broadcast Disk Paging," *Proc. of ACM STOC*, 1998.

27. Q. Hu, D. L. Lee and W-C. Lee, "Optimal Channel Allocation for Data Distrbution in Mobile Computing Environments," *Proc. of Intl. Conf. on Distributed Computing Systems (ICDCS)*, 1998.

28. W-C. Lee, Q. Hu and D. L. Lee, "A Study on Channel Allocation for Data Dissemination in Mobile Computing Environments," *Mobile Networks and Applications (MONET)*, Vol. 4, pp. 17–29, 1999.

29. G. Lee and S. C. Lo, "Broadcast Data Allocation for Efficient Access of Multiple Data Items in Mobile Environments," *Mobile Networks and Applications*, Vol. 8, pp. 365–375, 2003.

30. C-W. Lin, H. Hu and D-L. Lee, "Adaptive Realtime Bandwidth Allocation for Wireless Data Delivery," *ACM/Kluwer Wireless Networks (WINET)*, Vol. 10, pp. 103–120, 2004.

31. W. Ni, Q. Fang and S. V. Vrbsky, "A Lazy Data Approach for On-demand Data Broadcasting," *Proc. of 23rd Intl. Conf. on Distributed Computing Systems Workshops* (ICDCSW03).

32. JH. Oh, K. A. Hua and K. Prabhakara, "A New Broadcasting Technique for An Adaptive Hybrid Data Delivery in Wireless Mobile Network Environment," *Proc. of IEEE Intl. Performance Computing and Communications Conf.*, pp. 361–367, 2000.

33. W-C. Peng, J-L. Huang and M-S. Chen, "Dynamic Leveling: Adaptive Data Broadcasting in a Mobile Computing Environment," *ACM/KLUWER Mobile Networks and Applications*.

34. M. C. Pinotti and N. Saxena, "Push less and pull the current highest demanded data item to decrease the waiting time in asymmetric communication environments," *4th International Workshop on Distributed and Mobile Computing*, (IWDC), Calcutta, India, pp. 203–213. Springer-Verlag 2002; LNCS 2571, Dec 28–31, 2002.

35. N. Saxena, K. Basu and S. K. Das, "Design and Performance Analysis of a Dynamic Hybrid Scheduling for Asymmetric Environment," *IEEE Intl. Workshop on Mobile Adhoc Networks*, WMAN, 2004.

36. N. Saxena and M. C. Pinotti, "Performance Guarantee in a New Hybrid Push-Pull Scheduling Algorithm," *Third International Workshop on Wireless Information Systems (WIS)*, 2004.

37. N. Saxena, M. C. Pinotti and S. K. Das, "A Probabilistic Push-Pull Hybrid Scheduling Algorithm for Asymmetric Wireless Environment," *IEEE Intl. Workshop on Wireless Ad Hoc and Sensor Networks* (co-located with GlobeCom), 2004.

38. N. Saxena, K. Basu, S. K. Das and M. C. Pinotti, "A Prioritized Hybrid Scheduling for Two Different Classes of Clients in Asymmetric Wireless Networks," *24th IEEE. International Performance Computing and Communications Conference (IPCCC)*, 2005.

39. N. Saxena, K. Basu, S. K. Das and M. C. Pinotti, "A New Hybrid Scheduling Framework for Asymmetric Wireless Environments with Request Repetition," *3rd IEEE Intl. Symposium on Modeling and Optimization in Mobile, Ad Hoc, and Wireless Networks (WiOpt)*, 2005.

40. N. Saxena and M. C. Pinotti, "On-line Balanced K-Channel Data Allocation with Hybrid Schedule per Channel," *IEEE Intl. Conf. in Mobile Data Management (MDM)*, 2005.

41. N. Saxena, K. Basu, S. K. Das and M. C. Pinotti "A Dynamic Hybrid Scheduling Algorithm for Heterogeneous Asymmetric Environments," *International Journal of Parallel, Emergent and Distributes Systems (IJPEDS)*, Feb. 2005.

42. N. Saxena, K. Basu, S. K. Das and M. C. Pinotti, "A New Service Classification Strategy in Hybrid Scheduling to Support Differentiated QoS in Wireless Data Networks," *IEEE ICPP*, 2005.

43. N. Saxena and M. C. Pinotti, "A Dynamic Hybrid Scheduling Algorithm with Clients' Departure for Impatient Clients in Heterogeneous Environments," *IEEE International Workshop on Algorithms for Wireless, Mobile, Ad Hoc and Sensor Networks (WMAN)* 2005.

44. A. Seifert and M. H. Scholl, "Processing Read-Only Transactions in Hybrid Data Delivery Environments with Consistency and Currency Guarantees," *ACM/Kluwer Mobile Networks and Applications (MONET)*, Vol. 8, pp. 327–342, 2003.

45. C-J. Su, L. Tassiulas and V. J. Tsotras, "Broadcast Scheduling for Information Distribution," *ACM/Kluwer Wireless networks* (WINET), Vol. 5, pp. 137–147, 1999.

46. W. Sun, W. Su and B. Shi, "A Cost-Efficient Scheduling Algorithm of On-demand Broadcasts," *ACM / Kluwer Wireless Networks* (WINET), 2003.

47. "Support of Third Generation Services Using UMTS in a Converging Network Environment," *UMTS Forum*, 2002.

48. N. Vaidya and S. Hameed, Log time algorithms for scheduling single and multiple channel data broadcast, *Proc. Third ACM-IEEE Conf. on Mobile Computing and Networking (MOBICOM)*, pp. 90–99, September 1997.

49. N. H. Vaidya and S. Hameed, "Scheduling data broadcast in asymmetric communication environments," *Wireless Networks*, Vol. 5, pages 171–182, 1999.

50. N. Vlajic, C. C. Charalambous and D. Makrakis, "Performance Aspects of Data Broadcast in Wireless Networks with User Retrials," *IEEE Transactions on Networking*, Vol. 12, No. 4, pp. 620–633, 2004.

51. J. Xu, D. L. Lee, Q. Hu and W. C. Lee, *Data Broadcast: Handbook of Wireless Networks and Mobile Computing*, I. Stojmenovic, Ed., New York: Wiley, 2003.

52. P. Xuan, S. Sen, O. Gonzalez, J. Fernandez and K. Ramamritham, "Broadcast on Demand: Efficient and Timely Dissemination of Data in Mobile Environments," 1997.

53. W. G. Yee, S. B. Navathe, E. Omiecinski and C. Jermaine, "Efficient Data Allocation over Multiple Channels at Broadcast Servers," *IEEE Trans. on Computers*, Vol. 51, No. 10, pp. 1231–1236, 2002.

54. W. G. Yee and S. B. Navathe, "Efficient Data Access to Multi-channel Broadcast Programs," *Proc. of ACM Conf. on Intl. Knowledge Management, (CIIKM)*, pp. 153–160, 2003.

Index

A

AAT, 151
Access, 1, 6–10, 13, 15–34, 37–44, 47–63,
    69–73, 77–81, 87–95, 97–105,
    125–135, 137
  probability, 9, 13, 17, 18, 37, 42, 48–51, 65,
    67, 75, 80, 81, 87, 97, 101, 110, 137
  set, 7, 8, 37, 38
  time, 1, 6, 7, 10, 15, 19–25, 28–32, 37–40,
    56–61, 69–72, 87–95, 103–105,
    116, 125
ACR, 20, 27, 28
AED, 11, 123, 125–126
AMAX, 19, 26
Analytical, 35, 44, 50, 61, 62, 66, 71, 72, 89,
    91, 94, 95, 104, 105, 112, 122
  results, 61, 62, 71, 89, 91, 94–95, 104,
    105, 122
Anomaly, 13, 77, 81, 85, 91, 95
Asymmetric, 1, 2, 9, 11–12, 15–16, 37–38, 63,
    128, 137
Asymptotically optimal, 23
Average waiting time, 3, 24, 45, 53–54, 69, 73,
    84, 103, 138

B

Balking, 9, 76, 81–83
Batching, 20, 31
Bell-shaped, 57, 88, 91
Birth and death process, 14, 51, 67, 85, 137
Blackberry, 10, 108
Blocking, 1, 10, 13, 107, 110
BoD, 30
BOT, 35
Broadcast, 1–8, 11–12, 15–35, 38–44, 48–50,
    59, 72–73, 75, 77, 79, 87, 89, 97, 99,
    107–109, 116–117, 123–130, 137–138
Bucketing, 21

C

Caching, 25, 31, 138
Chapman-Kolmogrov's equation, 52, 68, 83,
Churning, 11, 116, 122
Client, 1–18, 22–35, 39–44, 48–54, 58, 65, 67,
    72–82, 85–95, 97–99, 106–112,
    116–117, 122–130, 133, 137–138
Cobham's iterative induction, 55, 115
Cold items, 24, 131

Competitive, 11, 18–19, 127

Competitive, 11, 18–19, 127
Cutoff point, 7, 9, 12–14, 39, 42–50, 56,
    61–63, 65, 70–73, 79–81, 88–95,
    104–107, 110–112, 114–122

D

Data access graph, 22
Data staging, 27
DBIS, 34
Delay, 1, 10–11, 13, 17–20, 28, 35, 41, 50, 56,
    63, 80, 107, 110, 112, 114–120, 123,
    135, 138
Departure, 14, 75, 77, 80–81, 87–89, 95,
    98, 102
Differentiated, 9, 10, 14, 84, 107–111, 113,
    115–116, 119, 121, 138
  QoS, 9, 10, 14, 107, 111–115, 122
  services, 9, 116
Dilation factor, 40
DirecPC, 8, 49
Dissemination, 1–3, 7, 12, 15, 25, 28–34
Downlink, 1–2, 6, 11, 15, 25, 30, 37–41,
    48, 108
Downstream, 1, 15016, 75
Drop requests, 13, 78, 82, 89, 138

E

ECC, 21
EDF, 9, 17, 16–33
Exponential, 23, 51, 67, 80–83, 99, 110

F

Fan out, 24, 26
FCFS, 19, 25–26, 40–41
Fibonacci, 23
FIFO, 30
Flat, 3, 11, 14–18, 59, 75–79, 84, 107–111,
    124–130, 138

G

Gaussian distribution, 32
Golden ratio, 23
Gray algorithm, 18
Greedy, 20–23, 33–34
Group-Id, 33
GSM, 8

H

HAT, 125
HDD, 31
Heterogeneous, 12–15, 25–26, 37, 45, 48, 63,
        85, 97–98, 107, 116, 137
Heuristics, 22
Hot items, 7, 17, 24, 37, 131,
Hybrid, 3, 7–15, 19–20, 29–35; 37–63,
        65–73, 75–94, 97–105, 107–121,
        123–134, 137
    waiting time, 72–731

I

Impatient, 1, 9, 13, 72, 75–81, 85, 95,
        98–99, 137
Importance factor, 10, 13, 107, 109,
        112–117, 138
Index, 17, 23, 28–29, 32, 40, 50–51, 72,
        99, 130
Interleaved, 14, 123

J

Jitter, 15, 19, 23–24, 35

L

Lazy, 18–19, 29
LDCF, 28
Little's formula, 54, 69, 84, 87, 114
Load, 14, 18, 28, 29, 31–32, 38, 40–42, 131,
        133, 137
Loan, 19, 33–34
Lossy, 21
Lower bound, 19, 21, 125, 128
LRU, 19, 27, 34
    LRU-LH, 27
LS, 27–28, 33
    LSAF, 33
LWF, 25–26, 28
LWTF, 25

M

MAED, 11, 125–135
$M/M/c/n$, 31
Markov, 14, 97–103, 137
Max Request Min Service Time, 12–13, 59,
        75, 97–98, 107
Max-heap, 43, 129
Memoryless, 61–122
MFA, 32
Modelling, 23, 37, 85, 122, 137–138
MRF, 5, 12, 14, 19, 25–26, 37, 41, 45, 59,
        63–67, 129, 137

MSS, 31
Multi 2, 11–12, 16–18, 22, 126–133, 138
    multichannel, 11–12, 126–133, 138
    multidisk, 16–18
    multivertex, 22
    multicast, 2

N

Non preemptive, 26
Nonpopular, 24–30
Normal mode, 27
Normalization, 53, 69
NP-hard, 22, 26

O

Occupancy, 53, 59, 69, 84, 86, 111–114
Opportunistic mode, 27,
Optimal, 1, 5, 7, 12, 15, 21–27, 31–37, 42–43,
        48, 56, 59–63, 88–91, 97–98, 104,
        107–108, 110–114, 125–126, 135, 137

P

Packet Fair Scheduling, 12
Paging, 18
PCS, 9, 138
Performance 1, 3, 6–14, 16–27, 34–38, 50–63,
        65–74, 80–89, 97–99, 103–107,
        110–116, 125–135, 137
    analysis, 14, 37, 51, 61, 65, 71, 81, 85, 99,
        104, 110, 116, 137
    evaluation, 66, 71, 110,
PFS, 3, 12, 20, 42–43, 48, 65–69, 73, 97–99,
        104, 137
Piggy back, 32
Point-to-point, 30–32
Poisson distribution, 21, 28, 51, 55, 67, 70, 80,
        82, 99, 109–110, 115, 131,
Polynomial 11, 15, 18–19, 20, 23, 56
    fit, 56
    time, 18
Popular, 1, 5–9, 13, 15–19, 34–33, 37, 50, 59,
        65, 75, 77, 80, 85, 89, 95, 97, 107,
        122, 137
Popularity, 1, 11–13, 18–20, 28, 32–33, 37, 50,
        65, 80, 95, 97, 122,
Pre-fetch, 16, 18, 25
Preemption, 26, 112
Prioritized cost, 116–121
Priority, 5, 10, 13, 28, 33, 49–58, 107–122
    scheduling, 107–108
Probabilistically, 12, 65–66, 70,
Probability, 9, 11, 13, 17–23, 37–42, 48–53,
        65–69, 75, 80–87, 97, 100–103,
        110–113, 123–126, 137

Pruning, 26–27
pseudo-popular, 77, 85
Pull, 5–10, 25–35, 37–58, 65–70, 75–88,
      97–105, 107–117, 123–131
   queue, 5, 9, 12, 28, 30, 39–54, 58–60,
      66–70, 75–85, 97–101, 107–112, 124,
      129, 137
Push, 5–10, 15–24, 37–58, 65–70, 75–88,
      97–105, 107–117, 123–131

Q

QoS, 1, 9–10, 13–14, 71–73, 107–122, 138
Queuing, 31, 80

R

Randomized algorithm, 18, 20, 23
Reneging, 9, 35, 76, 81–83,
Repeat attempt, 97–106.
Request set, 7–8, 37–39
Retrial, 14, 35, 97–99, 106
RTL, 28
RxW, 19, 26–27

S

Scheduling, 2–11, 20–34, 37, 43–50, 65–74,
      75–91, 106–91, 97–103, 107–110,
      123–127
Selection factor, 29
Sequential, 12, 65, 72–74
Service 9–14, 26, 45–71, 80, 97–98, 107–122,
      131–137,
   class, 9, 14, 107–122
   providers, 10–13, 107, 116, 122, 138,
   time, 12–13, 26, 45,–50, 54–71, 80, 97–98,
      107, 110, 114–115, 131, 137
Simulation, 45–58, 69–70, 88–95, 103–105,
      116–121, 131–135

Skew, 16–18, 23, 27, 37, 40, 44, 56–62, 65–68,
      60–75, 87–95, 97, 103–105, 110,
      116–117, 126–135
SLA, 10, 13, 27–28, 107
Spurious, 9, 14, 75–85, 95, 137
SRR, 121–128, 133–135
ST-MVSG, 35
STF, 25
Stretch, 5, 12–13, 19, 25–26, 37, 48–54, 59,
      63, 66, 75–79, 97–99, 104, 107–112,
      117, 137
   optimal, 5, 12, 37, 48, 63, 104, 107–108, 137

T

TC-AHB, 33
Tolerance, 11, 13, 72, 75, 77
Tuning time, 19, 28, 32

U

UMTS, 8
Unfairness, 107
Unicast, 2, 32, 35
Unique requests, 77–78, 108
Unit-length, 9, 14, 37
Uplink, 1–2, 6–7, 14–15, 25, 29–30, 37–39,
      123, 128
Upstream, 1, 15

W

Worst case, 23, 25, 29

Z

z-transform, 52–53, 68–69, 83–86, 111–113,
Zipf distribution, 23, 32, 37, 50, 65, 67, 80, 97,
      110, 117, 128, 131, 133,